T0400772

PUBLIC HEALTH IN THE 21ST CENTURY

THREAD THREAT

FORMALDEHYDE IN TEXTILES

PUBLIC HEALTH IN THE 21ST CENTURY

Additional books in this series can be found on Nova's website
under the Series tab.

Additional E-books in this series can be found on Nova's website
under the E-books tab.

CHEMISTRY RESEARCH AND APPLICATIONS

Additional books in this series can be found on Nova's website
under the Series tab.

Additional E-books in this series can be found on Nova's website
under the E-books tab.

PUBLIC HEALTH IN THE 21ST CENTURY

THREAD THREAT

FORMALDEHYDE IN TEXTILES

VICTORIA C. MUÑOZ
EDITOR

Nova Science Publishers, Inc.
New York

For permission to use material from this book please contact us:
Telephone 631-231-7269; Fax 631-231-8175
Web Site: http://www.novapublishers.com

Additional color graphics may be available in the e-book version of this book.

Library of Congress Cataloging-in-Publication Data

Thread threat : formaldehyde in textiles / editor, Victoria C. Muqoz.
p. cm.
Includes index.
ISBN 978-1-61324-839-3 (hardcover)
1. Formaldehyde--Health aspects. 2. Textile finishing agents--Health aspects. 3. Product safety. I. Muqoz, Victoria C.
RA766.F6T47 2011
615.9'5136--dc23

2011017608

Published by Nova Science Publishers, Inc. † New York

CONTENTS

PREFACE

Formaldehyde, one of the most widely produced chemicals in the world, is used in many products, including disinfectants, pressed wood and clothing and other textiles. Exposure to this chemical, which has been linked to adverse health effects for more than 30 years, typically occurs through inhalation and dermal contact. Formaldehyde can be used to enhance wrinkle resistance in some clothing and textiles, especially those made of cotton. The Consumer Product Safety Commission reviewed formaldehyde in clothing in the 1980s and determined that the levels found did not pose a public health concern. At that time, most clothing sold in the U.S. was made here, but the market has changed such that most U.S. clothing is now made in other countries. This book examines this market change which has raised new questions about the levels and dangers of formaldehyde in apparel, textiles and other consumer products.

Chapter 1-Formaldehyde—one of the most widely produced chemicals in the world—is used in many products, including disinfectants, pressed-wood, and clothing and other textiles. Exposure to this chemical, which has been linked to adverse health effects for more than 30 years, typically occurs through inhalation and dermal (skin) contact. Formaldehyde can be used to enhance wrinkle resistance in some clothing and textiles, especially those made of cotton. The Consumer Product Safety Commission reviewed formaldehyde in clothing in the 1980s and determined that the levels found did not pose a public health concern. At that time, most clothing sold in the United States was made here—but the market has changed such that most U.S. clothing is now made in other countries. This market change has raised new questions about the levels of formaldehyde in clothing.

Chapter 2- Formaldehyde is a toxic, pungent, water-soluble gas used in the aqueous form as a disinfectant, fixative, or tissue preservative, making it versatile for a wide range of uses. Formaldehyde resins are used in wood products (e.g. particleboard, paper towels), plastics, paints, manmade fibers (e.g. carpets, polyester), cosmetics, and other consumer products,[1] including many with which children have regular contact.[2] According to recent research and media reports, formaldehyde may be found in fabrics and children's clothing[3], children's furniture,[4] baby bath products,[5] and other products. Formaldehyde is also used in the resins used to bond laminated wood products and to bind wood chips in particleboard. Particleboard may be used in various types of furniture, including cribs and other items meant for use by or with children. The experience of Gulf Coast families living in mobile homes and travel trailers after Hurricane Katrina brought these hazards to the nation's attention; trailers, which have small, enclosed spaces, low air exchange rates, and many particleboard furnishings, may have much higher concentrations of formaldehyde than other types of homes.[6,7]

Chapter 3- In the summer of 2007 reports surfaced about high levels of lead in toys and other consumer goods and there were hundreds of thousands of items recalled. One area that initially escaped scrutiny at that time was textile and apparel product safety. Years before, the federal government recognized the lethal toxicity of asbestos fibers and TRIS flame retardant in children's sleepwear and acted appropriately to ban their use in consumer products. Today, once again, the question of safety is front and center and researchers are looking for answers regarding the safety of textiles and apparel. By researching the prevalence of other potentially toxic chemicals, such as formaldehyde, dyes and finishes, used every day in clothing, we will be able to determine just what chemicals and at what levels could pose risks to all of us, especially our children — and possibly lead to medical conditions ranging from contact dermatitis to neurotoxicity, endocrine disruption and possibly cancer.

Chapter 4- There have been no valid safety related problems raised in the US concerning the low levels of formaldehyde on clothing and textiles. In view of all the studies over the last 30 years indicting that there is not a formaldehyde problem with US textiles products and regulations already in place concerning formaldehyde and textiles, no new regulations are necessary. Because the evidence is so strong that formaldehyde in textiles does not pose a problem to consumers, there is no need for legislative or regulatory action concerning formaldehyde and textiles unless the results of the GAO study, required by Section 234 of the CPSIA which became law August 14, 2008, indicate that action is necessary.

Chapter 5- Persons exposed only to formaldehyde vapor do not pose substantial risks of secondary contamination. Persons whose clothing or skin is contaminated with a solution of formaldehyde can cause secondary contamination by direct contact or through off-gassing vapor.

- Formaldehyde is a colorless, highly toxic, and flammable gas at room temperature that is slightly heavier than air. It has a pungent, highly irritating odor that is detectable at low concentrations, but may not provide adequate warning of hazardous concentrations for sensitized persons.
- It is used most often in an aqueous solution stabilized with methanol (formalin).

Most formaldehyde exposures occur by inhalation or by skin or eye contact. Formaldehyde is absorbed well by the lungs, gastrointestinal tract, and, to a lesser extent, skin.

In: Thread Threat: Formaldehyde in Textiles ISBN: 978-1-61324-839-3
Editor: Victoria C. Muñoz

Chapter 1

FORMALDEHYDE IN TEXTILES: WHILE LEVELS IN CLOTHING GENERALLY APPEAR TO BE LOW, ALLERGIC CONTACT DERMATITIS IS A HEALTH ISSUE FOR SOME PEOPLE*

United States Government Accountability Office

WHY GAO DID THIS STUDY

Formaldehyde—one of the most widely produced chemicals in the world—is used in many products, including disinfectants, pressed-wood, and clothing and other textiles. Exposure to this chemical, which has been linked to adverse health effects for more than 30 years, typically occurs through inhalation and dermal (skin) contact. Formaldehyde can be used to enhance wrinkle resistance in some clothing and textiles, especially those made of cotton. The Consumer Product Safety Commission reviewed formaldehyde in clothing in the 1980s and determined that the levels found did not pose a public health concern. At that time, most clothing sold in the United States was made here—but the market has changed such that most U.S. clothing is

* This is an edited, reformatted and augmented version of a United States Government Accountability Office publication, Report to Congressional Committees GAO-10-875, dated August 2010.

now made in other countries. This market change has raised anew questions about the levels of formaldehyde in clothing.

In response to a mandate in the Consumer Product Safety Improvement Act of 2008, this report provides information on what is known about (1) the health risks of exposure to formaldehyde, particularly from clothing, and (2) the levels of formaldehyde in clothing sold in the United States. GAO analyzed government reviews and the medical literature, as well as studies on levels of formaldehyde in clothing, and had a sample of 180 textiles—primarily clothing—tested for formaldehyde by an accredited laboratory. While illustrative of formaldehyde levels that may be found in clothing, the test results from GAO's sample cannot be projected to all clothing sold in the United States. This report contains no recommendations.

WHAT GAO FOUND

The potential health risks associated with formaldehyde vary, depending largely on the means of exposure (e.g., inhalation or dermal contact), the concentration of the formaldehyde, and the duration of exposure. Inhaled formaldehyde may cause such effects as nausea, exacerbation of asthma, and cellular changes that may lead to the development of tumors. In fact, comprehensive reviews by the Department of Health and Human Services, the Environmental Protection Agency, and the World Health Organization have found that chronic inhalation exposure to formaldehyde may cause cancer. However, the health risk of greatest concern associated with formaldehyde in clothing—*allergic* contact dermatitis—stems from dermal exposure. A form of eczema, allergic contact dermatitis affects the immune system and produces reactions characterized by rashes, blisters, and flaky, dry skin that can itch or burn. Another potential health effect from dermal exposure to formaldehyde—*irritant* contact dermatitis—is also a form of eczema and has similar symptoms; however, this condition does not affect the immune system. Avoiding clothing containing formaldehyde is typically effective at preventing allergic and irritant contact dermatitis and relieving symptoms, but doing so can be difficult as clothing labels do not identify items treated with or containing formaldehyde. Washing clothing before it is worn often reduces formaldehyde levels but is not always successful. In some cases, avoiding or relieving allergic contact dermatitis requires more drastic measures, such as taking medications with potentially serious side effects. Finally, consumers may also experience dermal exposure to formaldehyde by using some cosmetics and skin care products, such as shampoos and sunscreens that contain formaldehyde.

Comprehensive data on formaldehyde levels in clothing sold in the United States are not publicly available. While formaldehyde levels in clothing are not regulated in the United States, the apparel industry reports that 13 countries have laws or regulations that limit formaldehyde levels in clothing. Most of the 180 items GAO had tested had formaldehyde levels that were below the most stringent of these industry-identified regulatory limits. GAO's test results are similar to those of recent studies of formaldehyde levels in clothing by the European Union, New Zealand, and Australia—that is, most items were found to meet the most stringent limits. Moreover, government studies we reviewed showed a decline in the formaldehyde levels in clothing since the 1980s, and the levels reported in these studies are generally consistent with the decline in levels reported in the medical literature. This decline is associated with the development and use of low-formaldehyde technologies (resins) in manufacturing clothing, which has been encouraged by such factors as the identification of formaldehyde as a probable human carcinogen via inhalation; the promulgation of federal regulations protecting workers from inhalation exposure to formaldehyde; and limits on formaldehyde levels that some U.S. retailers have established for clothing they sell.

ABBREVIATIONS

AATCC test	American Association of Textile Chemists and Colorists 112 test
CPSC	Consumer Product Safety Commission
DMDHEU	dimethylol dihydroxy ethylene urea
EPA	Environmental Protection Agency
HHS	Department of Health and Human Services
IARC	International Agency for Research on Cancer
ISO	International Organization for Standardization
Japanese test	Japanese Law 112 test
OSHA	Occupational Safety and Health Administration
USDA	Department of Agriculture
WHO	World Health Organization

August 13, 2010
The Honorable Jay Rockefeller
Chairman

The Honorable Kay Bailey Hutchison
Ranking Member
Committee on Commerce, Science and Transportation
United States Senate

The Honorable Henry A. Waxman
Chairman

The Honorable Joe Barton
Ranking Member
Committee on Energy and Commerce
House of Representatives

Formaldehyde is a colorless, pungent-smelling chemical well known for its use as a preservative and disinfectant in laboratories and mortuaries but is also widely used in consumer products such as pressed-wood products, glues and adhesives, cosmetics, and clothing and other textiles. Some clothing—generally garments made of cotton and other natural fibers—is treated with resins[1] containing formaldehyde primarily to enhance wrinkle resistance. Formaldehyde is toxic and has been linked to serious adverse health effects, including cancer, and some federal agencies have regulations that limit human exposure, which occurs primarily through inhalation and dermal (skin) contact. Regarding inhalation exposure, the Environmental Protection Agency (EPA) regulates formaldehyde emissions to the ambient air from both industrial sources and vehicles under the Clean Air Act, and the Department of Labor's Occupational Safety and Health Administration (OSHA) has standards in place that limit employee exposure to formaldehyde emissions in the workplace. Regarding dermal exposure to textiles, formaldehyde levels in clothing and other textiles that come into contact with the skin are not regulated. The Consumer Product Safety Commission (CPSC), which is charged with protecting the public from unreasonable risks of serious injury or death from consumer products, including clothing, reviewed formaldehyde in clothing sold in the United States in the 1980s and found that formaldehyde levels were sufficiently low so as not to be a public health concern.[2]

At the time of the CPSC review, most textiles sold in the United States were also manufactured in the United States. However, the market for textiles has changed significantly in recent years, raising questions about what the current levels of formaldehyde in clothing are. Currently, most clothing sold in the United States is imported from other countries, particularly from China, Vietnam, and other countries in Asia, as well as from countries in Central America. In 2008, for example, nearly 35 percent of clothing imported into the United States was manufactured in China, up from about 6 percent in 2000. Moreover, in contrast with the United States, some countries, including China, have established legal limits on the levels of formaldehyde that clothing may contain. For example, China and Japan have similar restrictions on levels of formaldehyde that may be contained in infant and other clothing that comes into direct contact with the skin. Further, the level of formaldehyde in clothing depends largely on the variability in the way the manufacturing process is conducted. For example, formaldehyde levels can vary among the same type and make of clothing because of, among other things, differences in the type of resin used and the ways it is applied.

Section 234 of the Consumer Product Safety Improvement Act of 2008 requires GAO to conduct a study on the use of formaldehyde in the manufacturing of textiles and apparel.[3] This report provides information on what is known about (1) the health risks from exposure to formaldehyde, particularly from clothing, and (2) the levels of formaldehyde found in clothing sold in the United States. To determine what is known about the health risks from exposure to formaldehyde, we analyzed comprehensive government reviews of the health effects of exposure to formaldehyde and conducted a literature review of articles in medical, textile, and environmental journals and books published from 1980 through April 2010 (see app. I for more information on this research).[4] To determine what is known about the levels of formaldehyde found in clothing sold in the United States, we analyzed information about (1) the use of formaldehyde-containing resins in clothing, (2) formaldehyde levels in clothing reported in government studies, and (3) relevant voluntary labeling programs, corporate limits, and regulatory limits for formaldehyde levels in clothing in the United States and other countries. Because of complexities in obtaining English translations of formaldehyde laws and regulations in other countries and confirming their application within the context of these countries' legal systems, we relied primarily on the American Apparel and Footwear Association for information on formaldehyde limits set in other countries, including the limits identified as the most stringent.[5] In addition, using an accredited commercial laboratory, we tested a nonprobability sample[6] of 180 textile items—primarily clothing (165) and some bed linens (15)—purchased from selected U.S. retailers in six metropolitan areas: Boston, Chicago, Dallas, Los Angeles, Seattle, and Washington, D.C. We selected low- or moderately priced items. We compared our test results with the most stringent regulatory limits for formaldehyde in clothing and other textiles established in other countries, as identified by the American Apparel and Footwear Association. The results from testing our nonprobability sample are illustrative of formaldehyde levels that may be found in some clothing and are not projectable to clothing sold in the United States in general. Throughout our review, we consulted with CPSC on our methodology and to obtain relevant background information stemming from the agency's mission to ensure consumer protection. Appendix II provides a more detailed description of our scope and methodology. We conducted this performance audit from September 2009 to August 2010 in accordance with generally accepted government auditing standards. Those standards require that we plan and perform the audit to obtain sufficient, appropriate evidence to provide a reasonable basis for our findings and conclusions based on our audit objectives. We believe that the evidence obtained provides a reasonable basis for our findings and conclusions based on our audit objectives.

BACKGROUND

Formaldehyde—one of the most widely produced and used chemicals in the world—is a water-soluble gas often used in a water-based solution (aqueous form) as a disinfectant or tissue preservative. In terms of toxicity, ingestion by an adult of as little as 1 ounce of a solution containing 37 percent formaldehyde has been reported to be lethal.[7] Formaldehyde is also used in other forms, including resins, that combine formaldehyde with other compounds. Aqueous formaldehyde and products containing formaldehyde tend to emit some formaldehyde into the air. Formaldehyde is also produced naturally in the environment and is found in low levels in people and most living things. In addition, formaldehyde is a by-product of combustion processes, such as wood burning and cigarette smoking. When formaldehyde is exposed to air, it begins to break down and dissipate.

The Department of Health and Human Services (HHS) reports that average levels of formaldehyde in outdoor air are often less than 0.01 parts of formaldehyde per million parts of air, and the average levels in indoor air in homes are often less than 0.04 parts per million.[8] Major sources of formaldehyde in outdoor air are man-made, such as power plants, manufacturing facilities, incinerators, and automobile exhaust. Formaldehyde levels in outdoor air are often found to be higher near some industry facilities and in heavily populated urban areas. In addition, people exposed to formaldehyde at work—such as medical personnel, embalmers, cabinetmakers, and textile plant employees—may be exposed to higher levels of formaldehyde. In general, the highest levels of airborne formaldehyde are detected indoors, where it can be released from various building materials, consumer products, and tobacco smoke.

Because formaldehyde is highly toxic, it is regulated by various federal and state agencies to protect human health and the environment.[9] For example, EPA lists formaldehyde as a hazardous air pollutant under the Clean Air Act and as a hazardous waste under the Resource Conservation and Recovery Act. Other agencies, such as OSHA and the Department of Housing and Urban Development, regulate airborne formaldehyde concentrations in the workplace and in manufactured homes, and the Food and Drug Administration limits the food-related use of formaldehyde to packaging components. In addition, California regulates formaldehyde emissions from composite wood products to protect human health from airborne exposure to formaldehyde.[10] According to an official from the California Environmental Protection Agency, the agency found that emissions from textiles commonly found in the home, such as draperies, dissipate quickly, whereas emissions from composite wood products are higher and remain relatively constant over time.

Formaldehyde-containing resins have been used in clothing and other textiles since the mid-1920s primarily to impart durable press characteristics to fabrics made from natural fibers, especially plant-based fibers such as cotton.[11] These resins may also provide other easy-care benefits, such as shrink resistance and color fastness.[12] The use of these resins in cotton clothing and other textiles became more prevalent in the 1950s and 1960s to compete with the increased use of synthetic fabrics, which often provided durable press characteristics. Under certain conditions, formaldehyde-containing resins may chemically degrade and release (off gas) free formaldehyde, including when exposed to high temperature and humidity.

The level of formaldehyde in clothing identified by testing and the level of formaldehyde that may be emitted by that clothing into the air will vary under different conditions—especially heat and humidity—and the test method used. Formaldehyde release mechanisms are numerous and complex, and emissions of formaldehyde from fabrics are much lower than the levels found in the fabrics by testing. Several analytical tests may be used to identify the levels of formaldehyde in clothing and other textiles. Test results are generally expressed as micrograms of formaldehyde per gram of fabric—typically either as microgram per gram or as parts per million.

The two analytical tests now commonly used in the textile industry to identify levels of formaldehyde in clothing and textiles are the Japanese Industrial Standard L 1041 test, also known as the Japanese Law 112 test (Japanese test), and the American Association of Textile Chemists and Colorists 112 test (AATCC test).[13] The Japanese test was developed to measure the amount of formaldehyde that may be released by clothing and other textiles that may come into contact with the skin, and the AATCC test was developed to measure the amount of formaldehyde that may be released from clothing and other textiles during extended storage or hot and humid conditions.[14] Because of different testing specifications, as well as variables related to the particular formaldehyde resins used, formaldehyde levels measured by the two tests are not comparable, and the results from the Japanese test cannot be used to predict what the results would be under the AATCC test and vice versa.[15] Regarding the tests performed on our sample of clothing and bed linens, the laboratory tested all of them using the Japanese test and a subset of them using the AATCC test. According to the American Apparel and Footwear Association, the Japanese test is used to determine whether formaldehyde levels in clothing are consistent with levels cited in international regulations. In addition, most of the U.S. retailers that provided us with information on internal corporate limits use the Japanese test or its equivalent. Further, the Japanese test was more frequently used in the studies we identified that conducted formaldehyde testing in clothing and other textiles.

The American Apparel and Footwear Association has identified the most stringent regulatory limits for formaldehyde in clothing and home textiles in other countries for its members that may wish to sell their products internationally.[16] The most stringent formaldehyde limits identified use the Japanese (or equivalent) test as a basis for measurement and are

- not detectable (defined as less than 20 parts per million) for products intended for children younger than 3 years of age,
- less than 75 parts per million for products that come into direct contact with the skin for children who are 3 years of age and older and for adults, and

- less than 300 parts per million for products that do not come into direct contact with the skin—e.g., outerwear—for children who are 3 years of age and older and for adults.

Some countries do not limit formaldehyde levels in clothing but require disclosure in labels if formaldehyde levels exceed specified amounts.[17] Further, some countries and private entities offer "eco labels" for clothing and other textiles, if their formaldehyde levels—as well as levels of other chemicals—are within specified ranges. Appendix III provides more information on voluntary labeling programs. Finally, while the United States does not legally limit formaldehyde levels in clothing, some U.S. retailers have set internal corporate limits on formaldehyde in clothing.

In the mid-1980s, most of the clothing sold in the United States was also manufactured in the United States and its territories. However, imports of clothing and other textiles from other countries now make up a majority of U.S. sales. Although textile industries are dispersed throughout the world, China is now the world's largest producer and exporter of clothing and textiles. Much of China's growth occurred during the 10-year phaseout of textile quotas under the 1995 World Trade Organization Agreement on Textiles and Clothing, which was completed on January 1, 2005.[18] As of 2008, China accounted for the largest share of total U.S. clothing imports— 34.3 percent, an increase from 6.5 percent in 1999. Vietnam and Bangladesh rank second and third at 6.7 percent and 6.3 percent, respectively. Appendix IV provides additional information on the 10 countries that exported the most clothing to the United States in 2008.

Formaldehyde Poses Different Health Risks Depending on the Type and Extent of Exposure

Key government reviews on the health effects of exposure to formaldehyde, including those conducted by EPA, HHS, and the World Health Organization (WHO), have concluded that chronic inhalation exposure to formaldehyde may cause cancer. Regarding exposure to formaldehyde in clothing, the health risk of greatest concern identified in key government reviews and in the medical literature is allergic contact dermatitis.

Key Government Reviews Have Concluded that Chronic Inhalation Exposure to Formaldehyde May Cause Cancer

The potential health risks from exposure to formaldehyde vary depending on the means of exposure (inhalation, dermal, ingestion, or eye contact), the concentration of formaldehyde, and the duration of exposure, among other factors. Inhaled formaldehyde may cause such effects as (1) discomfort or nausea stemming from the chemical's pungent odor; (2) irritation of the eyes, nose, and throat; (3) exacerbation of asthma; and (4) changes at the cellular level that may lead to the development of tumors. In fact, several comprehensive government reviews of the health risks of exposure to formaldehyde have found that chronic inhalation exposure to formaldehyde may cause cancer.

Concerns about the health risks of exposure to formaldehyde were heightened in 1979 when the Chemical Industry Institute of Toxicology[19] reported that formaldehyde caused nasal cancer in laboratory rats. Since then, government and industry entities have extensively studied the potential human health risks of inhalation exposure to this commonly used chemical. Concerns about formaldehyde are based on its cancer-causing potential in humans as well as its irritant properties. Research efforts on inhalation exposure have focused on indoor air levels of formaldehyde, where concentrations are known to be higher, and in particular on exposure in occupational settings. These research efforts include long-term epidemiological studies conducted on workplace exposure. For example, the largest study to date was conducted by the National Cancer Institute, which has tracked close to 26,000 workers employed in 10 different formaldehyde-producing or -using plants since the 1960s; the latest update to this study was published in 2009. Similarly, HHS's National Institute for Occupational Safety and Health has studied about 11,000 textile workers exposed to formaldehyde in the workplace. These studies have suggested that formaldehyde exposure is associated with nasopharyngeal cancer and possibly with cancers of the hematopoietic and lymphatic systems, particularly myeloid leukemia.[20]

Based on these and other studies, at least three government entities—EPA, HHS, and WHO—have conducted comprehensive reviews of the health effects of formaldehyde. For example, in 2005, the HHS's National Toxicology Program[21] concluded that chronically inhaled formaldehyde is reasonably anticipated to be a human carcinogen; the agency is currently considering upgrading this designation to "known to be a human carcinogen."[22] Similarly, beginning in 1987, EPA classified inhaled formaldehyde as a probable human carcinogen[23] and in June 2010 released a revised draft assessment classifying it as a known human carcinogen.[24] The draft EPA assessment is undergoing a review by the National Academies,[25] one of several key steps the agency must take prior to finalizing the assessment. In addition, in 2006, WHO's International Agency for Research on Cancer (IARC) reclassified formaldehyde from "probably carcinogenic to humans," a classification based on its 1995 assessment, to "carcinogenic to humans."[26] For the 2006 classification, IARC found sufficient evidence that formaldehyde causes nasopharyngeal cancer in humans, limited evidence for cancer of the nasal cavity and paranasal sinuses, and strong but not sufficient evidence for leukemia. In 2009, IARC's Cancer Monograph Working Group determined on the basis of additional epidemiological studies that there was sufficient evidence to associate formaldehyde exposure with leukemia.[27] This finding will be published in an upcoming IARC monograph.

Most studies supporting a link between exposure to formaldehyde and an increased risk of cancer studied workers exposed to formaldehyde occupationally, rather than people exposed to formaldehyde in consumer products, although people may be exposed to formaldehyde in consumer goods, such as pressed-wood products and textiles, that may "off gas" formaldehyde indoors. While EPA is required to develop regulations establishing standards for emissions of formaldehyde from hardwood plywood, medium-density fiberboard, and particleboard under July 2010 amendments to the Toxic Substances Control Act, the comprehensive government studies we reviewed do not indicate that formaldehyde levels in clothing present an inhalation health risk to consumers.

The Health Risk of Greatest Concern Associated with Dermal Exposure to Formaldehyde in Clothing is Allergic Contact Dermatitis

Regarding exposure to formaldehyde in clothing, CPSC officials said that, on the basis of research the agency conducted in the 1980s,[28] (1) there were no data indicating that formaldehyde in clothing and other textiles could penetrate the skin and cause cancer and (2) levels of formaldehyde found in clothing did not pose an acute or chronic health hazard to consumers. As a result, CPSC concluded that a regulatory standard was not needed for levels of formaldehyde in clothing and other textiles or for emissions of formaldehyde from these items. Further, the United States has no other legal limit on the level of formaldehyde that may be found in clothing and other textiles.

Based on our review of the medical literature, the health risk of greatest concern associated with formaldehyde in clothing—*allergic* contact dermatitis—stems from dermal exposure.[29] A form of eczema, allergic contact dermatitis produces reactions characterized by rashes, discoloration (particularly redness), swelling, blisters, scaling, and flaky dry skin that can itch or burn. These reactions are often very painful and can last indefinitely if left untreated. The reactions can be exacerbated by heat, humidity, friction, and perspiration and are usually worse where clothing fits closely. In some cases, repeated scratching of the affected area can lead to patches of tough, leathery skin. Allergic contact dermatitis also affects the immune system. Another potential health effect from dermal exposure to formaldehyde in clothing—*irritant* contact dermatitis—is also a form of eczema and has similar symptoms; however, this condition does not affect the immune system.

Formaldehyde is classified as a "strong sensitizer"—a substance that can cause hypersensitivity through recurring or prolonged contact. According to the medical literature, people with allergic contact dermatitis caused by contact with formaldehyde in clothing have generally become hypersensitive to the chemical through previous exposure. The test used to determine whether an individual has allergic contact dermatitis does not identify the level of formaldehyde in clothing that would trigger this condition. Although the estimates vary widely, the medical literature suggests that the amount of formaldehyde in clothing needed to trigger an allergic contact dermatitis reaction in sensitized individuals can be as little as 30 parts per million. The amount of formaldehyde in clothing that would cause a reaction is an area needing further research, according to a 2009 medical journal article by experts on contact dermatitis.[30]

Some of the medical literature has estimated the number of people with allergic and irritant contact dermatitis caused by dermal exposure to formaldehyde. This literature focuses on subpopulations of patients— those with eczema—and therefore the results cannot be generalized to the rest of the population. For example, one study reported that 9.2 percent of patients suspected of having contact dermatitis tested positive to a diagnostic skin test—called a patch test—that applied a 1 percent formaldehyde solution to the skin to test for any dermal reaction. Other studies estimate that between 1.2 percent and 2.3 percent of people with eczema have dermatitis related to formaldehyde in their clothing. Some of the medical literature suggests that allergic contact dermatitis may be underreported because medical practitioners might not distinguish it from cases of dermatitis with other causes, because of differences in how patch testing is conducted to determine dermatitis, and because some individuals may attempt to treat the condition themselves without seeking medical attention or are reluctant to make the number of visits to the doctor necessary to make a reliable diagnosis.

Avoiding clothing containing formaldehyde is typically effective at preventing allergic or irritant contact dermatitis and at relieving symptoms, but doing so may be difficult because labels for clothing sold in the United States generally do not provide information on formaldehyde content or on whether the clothing item was treated with formaldehyde.

One dermatologist with whom we spoke published an article that identifies some clothing companies that report using little or no formaldehyde in their clothing.[31] Also, a number of medical articles recommend that patients allergic to formaldehyde wash articles of clothing before wearing them to reduce the level of formaldehyde to which they may be exposed. The success of this technique in limiting exposure to formaldehyde in clothing may vary, however. For example, studies conducted by the Department of Agriculture (USDA) Agricultural Research Service report that the effect of laundering on formaldehyde levels depends on the type of resin that is applied to the clothing as well as on other factors, such as the alkalinity and hardness of the water and whether bleach is used. Other studies report that while formaldehyde levels may decline initially after washing, the levels may start increasing again after multiple washes. One publication explains that while washing clothing does remove some formaldehyde, levels may increase again over time as the resin is broken down by, among other things, washing and ironing.[32] Additionally, certain clothing items, such as hats, are generally not washed before being worn.

In some cases, more drastic measures may be necessary to avoid allergic contact dermatitis. Some researchers suggest that some patients may need to change occupations or job responsibilities to avoid contact with formaldehyde-containing products. If wearing clothing with formaldehyde cannot be avoided, some researchers suggest wearing synthetic or silk undergarments to act as a barrier between the skin and the clothing. When avoidance is not feasible or does not sufficiently relieve symptoms for those with allergic contact dermatitis, other treatment options include strong oral or topical medications, including immunosuppressive agents that may have serious side effects. Topical medications should be used with caution because some of these medications may actually contain formaldehyde, which could potentially perpetuate or worsen the reaction. Finally, we note that consumers may also experience dermal exposure to formaldehyde by using some cosmetics and skin care products, such as some shampoos and sunscreens that contain formaldehyde.

WHILE COMPREHENSIVE DATA IS NOT AVAILABLE, RECENT STUDIES SUGGEST THAT FORMALDEHYDE LEVELS IN CLOTHING ARE GENERALLY LOW AND HAVE DECLINED OVER TIME

In the absence of U.S. regulation of formaldehyde levels in clothing, and associated compliance testing, comprehensive data on the levels of formaldehyde in clothing sold in the United States are not publicly available.[33] Tests conducted by an accredited commercial laboratory of 180 items we purchased in stores across the country indicate th at formaldehyde levels in most of the items are low or not detectable. Moreover, these test results indicate that the levels of formaldehyde found in most of these items would meet the most stringent regulatory standards set by other countries as identified by the American Apparel and Footwear Association: not detectable in clothing and other items for infants and toddlers younger than 3 years of age and less than 75 parts per million for clothing and other items that come into direct contact with the skin for adults and for children 3 years of age and older.[34] However, 10 of the items exceeded these limits, with formaldehyde levels ranging from 75.4 to 206.1 parts per million. As table 1 shows, nine of these items exceeded the limits for adults and for children 3 years of age and older, and one item—a sheet for a child's crib—exceeded the limit for infants and toddlers younger than 3 years of age as well as that for adults and for children 3 years of age and older. More than half of the items we had tested that exceeded these limits were labeled as having fabric performance characteristics related to durable press, which may indicate the use of resins that contain formaldehyde and can alert consumers who wish to avoid formaldehyde in clothing.[35] Both domestic and imported clothing and bed linens had formaldehyde levels that exceeded the limits identified by the apparel industry. Appendix V provides more complete information on the items we had tested.

In 2007, the European Union and New Zealand government conducted studies of formaldehyde levels in clothing using methodologies similar to ours and also found that most clothing items tested had levels of formaldehyde that were low or not detectable.[36,37] Specifically, the European Union found that 212 of 221 items tested either had formaldehyde levels below 75 parts per million for adults and older children or levels that were not detectable for infants. Of the 9 items that exceeded these levels, 5 were either dress shirts labeled as "easy care" or T-shirts. Moreover, our analysis of the New Zealand government study showed that 96 of 99 items tested had formaldehyde levels below 75 parts per million; the 3 exceptions were men's dress or casual pants. A more limited study of 10 clothing items, conducted in 2007 by the Australian government, reported that all 10 items had formaldehyde levels that were not detectable.[38] These studies, as well as our own test results, provide important data on levels of formaldehyde that may be found in clothing, but these data cannot be projected to clothing in general.

Regarding the decline in levels of formaldehyde in clothing over time, government studies we reviewed have reported decreasing levels since the 1960s. While researchers have used various test methods in the past, which has limited comparisons of formaldehyde levels in clothing over time, the available data on formaldehyde levels in items tested show a decline over time. For example, the USDA Agricultural Research Service reported that in the early 1960s formaldehyde levels in clothing were found to be generally above 3,000 parts per million using the AATCC test.[39]

However, tests conducted in the early 1980s by the textile industry showed that formaldehyde levels in clothing were generally below 500 parts per million using the AATCC test. Further, tests conducted by government entities in Australia, Denmark, the European Union, Finland, New Zealand, and the United States from 1984 to 2010—more often using the Japanese test—show that the percentage of items with formaldehyde levels greater than 100 parts per million has generally declined (see table 2). For example, although in 1985 67 percent of items tested by CPSC[40] had levels of formaldehyde above 100 parts per million using the Japanese test, studies since 2003 have shown that 2 percent or less of the items tested using the Japanese test had formaldehyde levels above 100 parts per million. Similarly, the maximum formaldehyde level identified in any item in each study has generally declined.

Table 1. Ten Items Sold in the United States that Exceeded the Most Stringent Regulatory Standards for Formaldehyde Identified by the American Apparel and Footwear Association

Item type	Fiber content identified on label	Fabric performance characteristic identified on label or packaging	Country identified on label	Target customer	Formaldehyde level in parts per million[a]
Dress shirt	100% cotton	Wrinkle free	China	Men	206.1
Hat	100% cotton exclusive of decoration	None identified	China	Boys 3 years of age and older	192.6
Bed linens (pillow cases)	60% cotton, 40% polyester	Soft finish, easy care	Bahrain	Adults or children 3 years of age and older	189.6
Khakis	100% cotton	No iron, permanent crease	India	Men	169.6
Dress shirt	60% cotton, 40% polyester	None identified	China	Boys 3 years of age and older	95.1
Bed linens (pillow cases)	100% cotton	Wrinkle free, easy care, no ironing needed, eco-friendly	USA; fabric imported from Pakistan	Adults or children 3 years of age and older	93.8
Dress shirt	100% cotton	Noniron	Indonesia	Men	92.6
Bed linens (pillow cases)	100% cotton	Wrinkle free performance	Pakistan	Adults or children 3 years of age and older	89.3
Bed linens (crib sheet)	100% cotton	Preshrunk	Thailand	Infants/toddlers[b]	85.4
U.S. military combat uniform pants	50% cotton, 50% nylon	None identified	USA	Women	75.4

Source: GAO analysis of information provided on items' labels or packaging and test data from an accredited commercial laboratory.
[a]These formaldehyde levels were determined using the Japanese test. [b]Infants/toddlers refers to children younger than 3 years of age.

Table 2. Levels of Formaldehyde in Clothing and Other Textiles Reported in Government Studies, 1984-2010

Source	Year tested	Type of items tested	Number of items tested	Percentage of items tested with greater than 100 parts per million of formaldehyde		Maximum level of formaldehyde identified in any item (parts per million)	
				Japanese test[a]	AATCC test[a]	Japanese test[a]	AATCC test[a]
CPSC[b]	1984-1985	Clothing and bed linens	12	67	92	736.6	2,897
Tampere Regional Institute of Occupational Health, Finland	1986-1987	Clothing, home textiles, and cotton fabrics	20	50	90	855	1,680
Tampere Regional Institute of Occupational Health, Finland	1987-1994	Fabrics, textiles	144	11	c	2,000	c
Finnish Customs Laboratory	1988	Imported textiles	2,719	12	c	2,200	c
Finnish Customs Laboratory	1989	Imported textiles	1,922	7	c	1,050	c

Table 2. (Continued)

Source	Year tested	Type of items tested	Number of items tested	Percentage of items tested with greater than 100 parts per million of formaldehyde		Maximum level of formaldehyde identified in any item (parts per million)	
				Japanese test[a]	AATCC test[a]	Japanese test[a]	AATCC test[a]
Finnish Customs Laboratory	1990	Imported textiles	1,547	11	[c]	1,500	c
Finnish Customs Laboratory	1991	Imported textiles	2,173	9	[c]	805	c
Finnish Customs Laboratory	1992	Imported textiles	1,407	10	[c]	1,319	c
Finnish Customs Laboratory	1993	Imported textiles	1,680	5	[c]	643	c
Danish Ministry of Environment and Energy	2003[d]	Clothing and home textiles	10	0	[c]	82	c
New Zealand Ministry of Consumer Affairs	2007	Clothing, home textiles, and footwear	99	2	[c]	250	c

Source	Year tested	Type of items tested	Number of items tested	Percentage of items tested with greater than 100 parts per million of formaldehyde		Maximum level of formaldehyde identified in any item (parts per million)	
				Japanese test[a]	AATCC test[a]	Japanese test[a]	AATCC test[a]
Australian Competition and Consumer Commission	2007	Clothing	10	0	0	Not detectable	Not detectable
European Union	2007	Clothing and home textiles	221 items (Japanese test); 127 items (AATCC test)	1	9	162.5	397.3
GAO	2010	Clothing and bed linens	180 items (Japanese test); 21 items (AATCC test)	2	10	206.1	550.7

Sources: Government-sponsored or -supported studies and GAO analysis of test data from an accredited commercial laboratory

[a] Or an equivalent test method.

[b] CPSC-sponsored study tested a subset (12 items) of its full sample using the Japanese and AATCC tests. The entire sample of 180 clothing and bed linens was tested using a proprietary method developed by the Department of Energy's Oak Ridge National Laboratory, which conducted the testing. Based on this test method, 14 percent of the items had more than 100 parts per million of formaldehyde, and the maximum formaldehyde level identified in an item was 940.2 parts per million.

[c] Only the Japanese or Japanese-equivalent tests were used in this study. [d]This is the year the report was published.

The levels of formaldehyde found in the these studies are generally consistent with information reported since the 1980s in the medical literature and trade publications that documents a decline in formaldehyde levels in clothing since that time. This decline is associated with the development and use of low-formaldehyde resins in manufacturing clothing, which has been encouraged by such key factors as the following: (1) the identification of formaldehyde as a probable human carcinogen via inhalation and the promulgation of regulations by OSHA to protect workers, such as those in textile factories, from exposure to formaldehyde; (2) legal limits on the levels of formaldehyde in clothing and other textiles adopted in other countries; (3) limits on formaldehyde levels that some U.S. retailers have established for clothing they sell; and (4) the textile industry's development of improved resins that provide durable press characteristics but release less formaldehyde.

More specifically, as discussed earlier, formaldehyde was classified as a probable human carcinogen in the 1980s based on animal and human studies showing that airborne formaldehyde exposure is associated with certain types of cancer. In 1987 and 1992, OSHA decreased the permissible airborne exposure level of formaldehyde in the workplace. The current OSHA regulation, among other things, limits airborne exposure to 0.75 parts of formaldehyde per million parts of air.[41] While this regulation primarily addresses the airborne concentration of formaldehyde to which workers—such as those in textile factories—may be exposed, several government studies and the medical literature have noted that OSHA regulations have encouraged lower formaldehyde levels in clothing and other textiles. Information provided in a Finnish Regional Institute of Occupational Health study illustrates the relationship between the levels of formaldehyde being applied to textiles in the workplace and airborne formaldehyde emissions in the factory. The Finnish study reported that formaldehyde levels in textiles should be under 200 parts per million[42] to ensure airborne formaldehyde emission levels in factories remain below 1 part per million under adverse conditions such as low ventilation and high humidity.[43]

Another factor that may have encouraged the use of lower levels of formaldehyde in clothing is the adoption in other countries of legal limits on formaldehyde in clothing and textiles and on the airborne levels of formaldehyde in the workplace. For example, according to HHS's National Institute for Occupational Safety and Health, over 20 countries have established occupational exposure limits to protect workers, including those in the textile industry.[44] In addition, the European Union's European Scientific Committee on Occupational Exposure Limits has issued provisional occupational exposure limits for formaldehyde. Further, according to the American Apparel and Footwear Association, 13 countries regulate the level of formaldehyde in clothing.[45] Another factor that may limit formaldehyde levels in clothing is the use of voluntary labeling programs such as the Oeko-Tex® Standard 100 label.[46] Such labels may be displayed by clothing and other textile items that meet certain limits on formaldehyde and other substances. Appendix III provides more information on voluntary labeling programs.

In addition, while there are no legal limits on the levels of formaldehyde in clothing in the United States, some U.S. retailers have established their own corporate limits on formaldehyde in clothing and bed linens. These corporate formaldehyde limits are generally considered proprietary, although some may be disclosed upon request. We obtained information on corporate limits for formaldehyde in clothing established by 14 U.S.-based retailers.[47] Many of these limits were set within the last 10 years, although one was set as early as 1995. In some cases, the limits established by retailers apply only to private store brands and not to national brands, some of which may establish their own limits. Some of the 14 retailers told us they review regulations adopted in other countries to inform their own corporate limits or to comply with regulations in countries where they operate.

Table 3 shows the corporate limits for formaldehyde in clothing and bed linens sold in the United States that we obtained from 14 retailers. The retailers' limits vary both by age and item category. For example, a retailer may have limits for clothing and bed linens for infants and toddlers but not for clothing or bed linens for adults.

However, retailers may use their corporate limits more as guidelines than as absolute compliance limits. For example, one retailer told us that when its test results for formaldehyde content exceed corporate limits, the company may conduct additional testing to determine the extent of the problem to inform its response. In contrast, this retailer told us that if a single sample exceeds a U.S. regulatory standard—for example, for lead in clothing—the company rejects the entire shipment. Another retailer told us the company complies with Japan's formaldehyde limit—the most stringent—when selling products in Japan but has established a less stringent limit for clothing sold in the United States.

Table 3. Formaldehyde Limits Set by 14 U.S. Retailers for Clothing and Bed Linens

	Number of retailers reporting formaldehyde limits, by category					
Formaldehyde limits[a,b]	Clothing for infants/ toddlers	Bed linens for infants/ toddlers	Clothing for children	Bed linens for children	Clothing for adults	Bed linens for adults
None	0	1	0	1	3	3
<250 parts per million	0	1	0	1	0	1
<200 parts per million[c]	1	1	1	1	1	2
<150 parts per million	0	1	0	1	0	1c
<100 parts per million	0	0	1	1	2	1
<75 parts per million	2	4	11d	9d	8d	6c
<20 parts per million	11c	6[d]	1	0	0	0
Number of retailers reporting limits	14	14	14	14	14	14

Source: GAO analysis of data from 14 U.S. retailers.

Notes: Retailers' age designations for infants/toddlers and children vary.

[a] Unless otherwise noted, retailers reported using the Japanese test.

[b] Formaldehyde limits for clothing are applicable to clothing that comes in direct contact with the skin.

[c] One retailer determines compliance with this limit using the AATCC test.

[d] Two retailers determine compliance with this limit using the AATCC test.

Finally, another factor that may have contributed to lower levels of formaldehyde in clothing, according to government and trade publications, is industry actions to address concerns about some formaldehyde-containing resins. The older resins, such as urea formaldehyde and melamine formaldehyde, impart durable press characteristics to clothing but also tend to release more formaldehyde during the manufacture, storage, retailing, and use of fabrics and clothing than newer resins because they are less chemically stable. In addition, the older resins can also stiffen fabric, degrade after repeated washing, damage fabrics if chlorine bleach is used, and cause the fabrics to emit a noticeable odor.

The development and use of newer resins in clothing production to impart durable press characteristics have addressed some of these issues as well as reduced the level of formaldehyde in clothing. These newer resins, also called cross-linking agents, became widely used in the 1980s. For example, dimethylol dihydroxy ethylene urea (DMDHEU) and its derivatives are reported to be the most commonly used resins today. Fabrics finished with DMDHEU may release moderate amounts of formaldehyde but can be modified to release low to ultra-low levels.[48] Newer resins that do not contain formaldehyde have also been developed, but they are not as widely used because of their higher cost; some may also have negative effects on fabrics.

AGENCY COMMENTS AND OUR EVALUATION

We provided a draft of this report to the Chairman, CPSC, for review and comment. CPSC provided technical comments, which we have incorporated, as appropriate.

We are sending copies of this report to the appropriate congressional committees; the Chairman, CPSC; and other interested parties. The report is also available at no charge on the GAO Web site at http:www.gao.gov.

If you or your staff have any questions about this report, please contact me at (202) 512-3841 or stephensonj@gao.gov. Contact points for our Offices of Congressional Relations and Public Affairs may be found on the last page of this report. GAO staff who made major contributions to this report are listed in appendix VI.

John B. Stephenson
Director, Natural Resources
and Environment

Appendix I: Government Reviews of the General Health Effects of Exposure to Formaldehyde and Medical Literature on the Health Effects of Formaldehyde in Textiles

Following is more detailed information on the sources we used in our analysis to determine what is known about the health effects of exposure to formaldehyde.

Aalto-Korte, K., R. Jolanki, and T. Estlander. "Formaldehyde-Negative Allergic Contact Dermatitis from Melamine-Formaldehyde Resin." *Contact Dermatitis*, vol. 49, no. 4 (2003).

Andersen, Klaus E. and Howard I. Maibach. "Multiple Application Delayed Onset Contact Urticaria: Possible Relation to Certain Unusual Formalin and Textile Reactions?" *Contact Dermatitis*, vol. 10, no. 4 (1984).

Andersen, Klaus E. and Knud Hamann. "Cost Benefit of Patch Testing with Textile Finish Resins." *Contact Dermatitis*, vol. 8, no. 1 (1982).

Australia Department of Health and Ageing. *National Industrial Chemicals Notification and Assessment Scheme, Priority Existing Chemical Assessment Report No. 28, Formaldehyde*. Sydney, Australia, 2006.

Belsito, Donald V. "What's New in Contact Dermatitis: Textile Dermatitis." *American Journal of Contact Dermatitis*, vol. 4, no. 4 (1993).

Bracamonte, B.G., F.J. Ortiz de Frutos, and L.I. Diez. "Occupational Allergic Contact Dermatitis due to Formaldehyde and Textile Finish Resins." *Contact Dermatitis*, vol. 33, no. 2 (1995).

Carlson, Ryan M., Mary C. Smith, and Susan T. Nedorost. "Diagnosis and Treatment of Dermatitis Due to Formaldehyde Resins in Clothing." *Dermatitis*, vol. 15, no. 4 (2004).

Cockayne, Sarah E., Andrew J.G. McDonagh, and David J. Gawkrodger. "Occupational Allergic Contact Dermatitis from Formaldehyde Resin in Clothing." *Contact Dermatitis*, vol. 44, no. 2 (2001).

Cronin, Etain. *Contact Dermatitis*, New York City, New York: Churchill Livingstone, 1980.

De Groot, Anton C., et al. "Formaldehyde-Releasers: Relationship to Formaldehyde Contact Allergy. Formaldehyde-Releasers in Clothes: Durable Press Chemical Finishes. Part 1." *Contact Dermatitis*, vol. 62, no. 5 (2010).

De Groot, Anton C. and Freddy Gerkens. "Contact Urticaria from a Chemical Textile Finish." *Contact Dermatitis*, vol. 20, no. 1 (1989).

De Groot, Anton C. and Howard I. Maibach. "Does Allergic Contact Dermatitis from Formaldehyde in Clothes Treated with Durable-Press Chemical Finishes Exist in the USA?" *Contact Dermatitis*, vol. 62, no. 3 (2009).

Donovan, Jeff and Sandy Skotnicki-Grant. "Allergic Contact Dermatitis from Formaldehyde Textile Resins in Surgical Uniforms and Nonwoven Textile Masks." *Dermatitis*, vol. 18, no. 1 (2006).

Environment Canada and Health Canada, *Canadian Environmental Protection Act, 1999, Priority Substances List Assessment Report, Formaldehyde.* Canada, 2001.

Fowler, Joseph F. "Formaldehyde as a Textile Allergen." In *Textiles and the Skin (Current Problems in Dermatology)*, vol. 31, edited by P. Elsner, K. Hatch, and W. Wigger-Alberti, 156-165. Basel, Switzerland: S. Karger AG, 2003.

Fowler, Joseph F. Jr., Steven M. Skinner, and Donald V. Belsito. "Allergic Contact Dermatitis from Formaldehyde Resins in Permanent Press Clothing: An Underdiagnosed Cause of Generalized Dermatitis." *Journal of the American Academy of Dermatology*, vol. 27, no. 6 (1992).

Geldof, B.A., I.D. Roesyanto, and Th. van Joost. "Clinical Aspects of Para-Tertiary-Butylphenolformaldehyde Resin (PTBP-FR) Allergy." *Contact Dermatitis*, vol. 21, no. 5 (1989).

Hatch, Kathryn L. "Chemicals and Textiles Part II: Dermatological Problems Related to Finishes." *Textile Research Journal*, vol. 54, no. 11 (1984).

Hatch, Kathryn L. and Howard I. Maibach. "Textile Chemical Finish Dermatitis." *Contact Dermatitis*, vol. 14, no. 1 (1986).

Hatch, Kathryn L. and Howard I. Maibach. "Textile Dermatitis: An Update." *Contact Dermatitis*, vol. 32, no. 6 (1995).

Hatch, Kathryn L. and Howard I. Maibach. "Textile Dye Dermatitis." *Journal of the American Academy of Dermatology*, vol. 32, no. 4 (1995).

Hegewald, Janice, et al. "Meteorological Conditions and the Diagnosis of Occupationally Related Contact Sensitizations." *Scandinavian Journal of Work, Environment & Health*, vol. 34, no. 4 (2008).

Imbus, Harold R. "Clinical Evaluation of Patients with Complaints Related to Formaldehyde Exposure." *Journal of Allergy and Clinical Immunology*, vol. 76, no. 6 (1985).

Iversen, Olav Hilmar. "Formaldehyde and Skin Carcinogenesis." *Environment International*, vol. 12, no. 5 (1986).

Lazarov, A. "Textile Dermatitis in Patients with Contact Sensitization in Israel: a 4-Year Prospective Study." *Journal of the European Academy of Dermatology and Venereology*, vol. 18, no. 5 (2004).

Lazarov, A. and M. Cordoba. "Purpuric Contact Dermatitis in Patients with Allergic Reaction to Textile Dyes and Resins." *Journal of the European Academy of Dermatology and Venereology*, vol. 14, no. 2 (2000).

Le Coz, Christophe-J. "Clothing." In *Textbook of Contact Dermatitis*, 3rd ed., edited by R.J.G. Rycroft, T. Menne, P.J. Frosch, and Jean-Pierre Lepoittevin, 727-749. Berlin, Heidelberg, Germany: Springer, 2001.

Maibach, Howard. "Formaldehyde: Effects on Animal and Human Skin." In *Formaldehyde Toxicity*, edited by J.E. Gibson, 166-174. Washington, D.C.: Hemisphere, 1983.

Marks, James G. Jr., et al. "North American Contact Dermatitis Group Patch-Test Results, 1998 to 2000." *American Journal of Contact Dermatitis*, vol. 14, no. 2 (2003).

Metzler-Brenckle, Lorrie and Robert L. Rietschel. "Patch Testing for Permanent-Press Allergic Contact Dermatitis." *Contact Dermatitis*, vol. 46, no. 1 (2002).

National Academy of Sciences, Committee on Toxicology, Board of Toxicology and Environmental Health Hazards, Assembly of Life Sciences, National Research Council. *Formaldehyde—An Assessment of the Health Effects, Prepared for the Consumer Product Safety Commission*. Washington, D.C., 1980.

Nordman, Henrik, Helena Keskinen, and Matti Tuppurainen. "Formaldehyde Asthma—Rare or Overlooked?" *The Journal of Allergy and Clinical Immunology*, vol. 75, no. 1 (1985).

Organisation for Economic Co-operation and Development. *SIDS Initial Assessment Report for 14th SIAM, Formaldehyde (CAS No. 50-00-0)*. Paris, France: United Nations Environment Programme Publications, 2002.

Reich, Hilary C. and Erin M. Warshaw. "Allergic Contact Dermatitis from Formaldehyde Textile Resins." *Dermatitis*, vol. 21, no. 2 (2010).

Rietschel, Robert L. and Joseph F. Fowler, Jr. "Textile and Shoe Dermatitis." In *Fisher's Contact Dermatitis*, 4th ed., 358-413. Baltimore, Maryland: Williams & Wilkins, 1995.

Robbins, Joe D., et al. "Bioavailability in Rabbits of Formaldehyde from Durable-Press Textiles." *Journal of Toxicology and Environmental Health*, vol. 14, no. 2-3 (1984).

Scheman, Andrew J., et al. “Formaldehyde-Related Textile Allergy: An Update.” *Contact Dermatitis*, vol. 38, no. 6 (1998).

Seidenari, S., B.M. Manzini, and P. Danese. “Contact Sensitization to Textile Dyes: Description of 100 Subjects.” *Contact Dermatitis*, vol. 24, no. 4 (1991).

Seidenari, S., B.M. Manzini, P. Danese, and A. Motolese. “Patch and Prick Test Study of 593 Healthy Subjects.” *Contact Dermatitis*, vol. 23, no. 3 (1990).

Sherertz, Elizabeth F. “Clothing Dermatitis: Practical Aspects for the Clinician.” *American Journal of Contact Dermatitis*, vol. 3, no. 2 (1992).

Trattner, Akiva and Michael David. “Textile Contact Dermatitis Presenting as Lichen Amyloidosus.” *Contact Dermatitis*, vol. 42, no. 2 (2000).

Trattner, A., J.D. Johansen, and T. Menne. “Formaldehyde Concentration in Diagnostic Patch Testing: Comparison of 1% with 2%.” *Contact Dermatitis*, vol. 38, no. 1 (1998).

U.S. Department of Health and Human Services, Agency for Toxic Substances and Disease Registry. *Medical Management Guidelines for Formaldehyde*, http://www.atsdr.cdc.gov/mhmi/mmg111.html.

U.S. Department of Health and Human Services, Agency for Toxic Substances and Disease Registry. *Toxicological Profile for Formaldehyde*. Atlanta, GA, 1999.

U.S. Department of Health and Human Services, Public Health Service, National Toxicology Program. *Report on Carcinogens, 11th Edition*. Research Triangle Park, NC, 2005.

U.S. Department of Health and Human Services, Public Health Service, National Toxicology Program. *Report on Carcinogens, Background Document for Formaldehyde*. Research Triangle Park, NC, 2010.

U.S. Environmental Protection Agency. *Integrated Risk Information System, Formaldehyde (CASRN 50-00-0)*, http://www.epa.gov/ncea/iris/subst/0419.htm.

U.S. Environmental Protection Agency. IRIS *Toxicological Review of Formaldehyde—Inhalation Assessment (External Review Draft)*. Washington, D.C., June 2010.

Washington State Department of Labor and Industries, Safety and Health Assessment and Research for Prevention, *Clothing Dermatitis and Clothing-Related Skin Conditions*. Report: 55-8-2001.

World Health Organization. *Concise International Chemical Assessment Document 40, Formaldehyde*. Geneva, Switzerland: United Nations

Environment Programme, the International Labour Organization, and the World Health Organization, 2002.

World Health Organization, International Agency for Research on Cancer. *IARC Monographs on the Evaluation of Carcinogenic Risks to Humans, Volume 88, Formaldehyde, 2-Butoxyethanol and 1-tert-Butoxypropan-2-ol*. Lyon, France, 2006.

APPENDIX II: OBJECTIVES, SCOPE, AND METHODOLOGY

Section 234 of the Consumer Product Safety Improvement Act of 2008 requires GAO to conduct a study on the use of formaldehyde in the manufacturing of textiles and apparel.[1] This report provides information on what is known about (1) the health risks from exposure to formaldehyde, particularly from clothing, and (2) the levels of formaldehyde found in clothing sold in the United States. Throughout our review, we consulted with the Consumer Product Safety Commission (CPSC) on our methodology and to obtain relevant background information stemming from the agency's mission to ensure consumer protection.

To determine what is known about the health effects of exposure to formaldehyde, particularly from clothing, we conducted literature reviews on both the general health effects of formaldehyde and the health effects of exposure to formaldehyde in clothing. For the general health effects, we summarized the findings of several comprehensive government reviews, including (1) the Health and Human Services' (HHS) 2005 National Toxicology Program report on carcinogens (currently being updated),[2] (2) HHS's Agency for Toxic Substances and Disease Registry 1999 toxicological profile for formaldehyde,[3] (3) the Environmental Protection Agency's (EPA) 1991 Integrated Risk Information System formaldehyde assessment[4] and the 2010 draft update,[5] (4) the 2006 monograph on formaldehyde prepared by the World Health Organization's International Agency for Research on Cancer,[6] and (5) the Organisation for Economic

Co-operation and Development's 2002 Screening Information Data Set for formaldehyde.[7]

To determine the health effects of exposure to formaldehyde in clothing, we reviewed and summarized the findings of over 40 relevant articles published from 1980 to April 2010 primarily in medical journals, but also in textile industry and environmental journals and books. We targeted reviews and meta-analyses, rather than individual studies, published in English but also included articles presenting original research, such as clinical studies. In addition, we interviewed and obtained studies from medical professionals, officials from the Department of Agriculture (USDA) and CPSC, and representatives from industry organizations such as the American Apparel and Footwear Association, Cotton Incorporated, and the Formaldehyde Council, Inc.

To determine what is known about the levels of formaldehyde found in clothing sold in the United States, we analyzed information on formaldehyde-containing resins and formaldehyde levels in clothing from government studies, patent applications, and trade publications; interviewed officials from government agencies including USDA's Agricultural Research Service, industry associations, retailers, testing laboratories, and academia, as well the medical profession; and contracted with an accredited commercial laboratory to test the formaldehyde levels in a nonprobability sample[8] of 180 articles of clothing (165) and bed linens (15).

For our nonprobability sample, we purchased clothing and bed linens from 10 national retailers, two military facilities, and a store selling children's scout uniforms. In March 2010, our staff purchased items in six major metropolitan areas: Boston, Chicago, Dallas, Los Angeles, Seattle, and Washington, D.C. Our sample was approximately evenly split by gender and included items worn by adults and children. We purchased only items that come into direct contact with the skin; we did not buy outerwear such as jackets or coats. We used clothing sizes to determine whether clothing was intended for children 3 years of age and older. We also considered crib sheets and cloth diapers to be intended for use by infants and toddlers. The results from testing our nonprobability sample are illustrative of formaldehyde levels that may be found in some clothing and are not projectable to clothing sold in the United States in general.

We selected items that were likely to contain formaldehyde, such as those made from plant-based fibers, like cotton, that are likely to be treated for durable press. Almost all of the items in the sample were made from 50 percent or more natural (mostly plant-based) fibers. About 76 percent of the items were made from 90 percent or more cotton. About a fifth of the items were labeled as being treated for durable press, including bed linens and career apparel such as dress shirts. We also included some commonly worn items, such as T-shirts and jeans, based on a review of Department of Commerce data for volumes of clothing and bed linens imported or produced in the United States. We selected low- or moderately priced items. Most of the items in our sample were imported; however, we also made an effort to ensure that about 10 percent of them were American made.

An accredited commercial laboratory tested the items for formaldehyde. The laboratory is accredited by, among others, the American Association of Textile Chemists and Colorists and the American Association for Laboratory Accreditation.[9] To prepare items for shipment, we placed each one in its own zippered plastic bag shortly after its purchase to avoid cross-contamination from other items or packaging materials. Samples received by the laboratory were cut, weighed, and kept wrapped in foil inside a plastic bag until the formaldehyde test was run.

The laboratory tested the items using two test methods commonly used by the textile industry—the Japanese Industrial Standard L 1041 test, also known as the Japanese Law 112 test (Japanese test), and the American Association of Textile Chemists and Colorists 112 test (AATCC test). Results from the Japanese test and the AATCC test are not comparable because, for example, they use different testing specifications. The procedure the laboratory used for the Japanese test included placing 1 gram of the sample in a stoppered flask with 100 mL of water and immersing it in a bath of 40°C water for 1 hour. The procedure the laboratory used for the AATCC test included suspending 1 gram of the sample in a sealed jar above 50 mL of water. The jar was kept in an oven at 49°C for 20 hours and then cooled at room temperature for 30 minutes. For both tests, an acetyl acetone reagent was added as the next step, and the solution was heated at 40°C for 30 minutes and allowed to cool at room temperature for 30 minutes. The laboratory ran quality control tests for each batch of samples.

We had each of the 180 items tested using the Japanese test and a smaller number using the AATCC test because most of the regulatory and corporate formaldehyde limits we identified used the Japanese test or its equivalent. According to the American Apparel and Footwear Association, this test or its equivalent is used to determine whether formaldehyde levels in clothing are consistent with levels cited in international formaldehyde regulations, which we used as a basis of comparison for our own test results. Also, the Japanese test was more frequently used in the government studies we identified that conducted formaldehyde testing in clothing and other textiles. In addition, most of the U.S. retailers that provided us with information on internal corporate limits on formaldehyde in clothing, and the voluntary labeling programs we identified, use the Japanese test or its equivalent. We tested 21 of the 180 items using both the Japanese test and the AATCC test. This latter test is used by some U.S. retailers that have established corporate limits on formaldehyde and has been used in some studies cited in government reviews, trade publications, and the medical literature. The 21 items tested by both methods were selected to include a similar mix of clothing and bed linens as the overall sample.

If items contained more than one type of fabric that could potentially contain varying levels of formaldehyde, the laboratory tested a composite sample that included equal amounts of these fabrics. Decorative elements that do not come into direct skin contact, such as sequins, were not included as part of the composite. This composite approach would tend to result in formaldehyde levels falling between the highest and lowest levels in the different fabrics.

In addition, we contacted U.S.-based, publicly owned retailers to determine if they have internal corporate limits for formaldehyde in clothing. We contacted 16 retailers and received responses from 14, of which 13 had limits. We also found information on the Internet on the limits set by one of the two retailers that did not respond.

To identify regulatory and voluntary labeling programs that may be used in other countries as well as relevant studies on the levels of formaldehyde in clothing, we contacted government and European Union officials and conducted Internet searches. We identified a number of countries that have regulatory standards for formaldehyde in textiles as well as voluntary labeling programs. Because of complexities in obtaining translations of other countries' formaldehyde laws and regulations into English and confirming their application within the context of these countries' legal systems, we relied primarily on the American Apparel and Footwear Association for information on formaldehyde limits set in other countries, including the limits identified as the most stringent.[10] We obtained information on international worker safety standards for formaldehyde primarily from a report by HHS's National Institute for Occupational Safety and Health.

We conducted this performance audit from September 2009 to August 2010 in accordance with generally accepted government auditing standards. Those standards require that we plan and perform the audit to obtain sufficient, appropriate evidence to provide a reasonable basis for our findings and conclusions based on our audit objectives. We believe that the evidence obtained provides a reasonable basis for our findings and conclusions based on our audit objectives.

Appendix III: Voluntary Labeling Programs

Table 4 lists examples of voluntary labeling programs, which are primarily used in other countries.[1] Clothing and other items meeting the specified limits for formaldehyde and other substances would typically display a label indicating their compliance. The formaldehyde limits in the table are those applicable to clothing and other textiles that come into direct contact with the skin.

Table 4. Examples of Voluntary Labeling Programs

Governing body	Label name	Formaldehyde limit for infants/toddlers,[a] in parts per million	Formaldehyde limit for children and adults, in parts per million	Test method specified (or equivalent)
European Union	European Ecolabel	20	30	Japanese test
Good Environmental Choice Australia Ltd	Australian Ecolabel	30	30	Japanese or AATCCtests[b]
International Association for Research and Testing in the Field of Textile Ecology	Oeko-Tex® Standard 100	Not detectable[c]	75	Japanese test
International Working Group on Global Organic Textile Standard	Global Organic Textile Standard	Any level prohibited	Any level prohibited	Not specified
New Zealand Ecolabelling Trust	Environmental Choice New Zealand	30	30	Japanese test

Source: GAO analysis of voluntary labeling programs.

[a] The transition from limits applicable to infants/toddlers to those for children and adults, where applicable, is 3 years of age.

[b] Or certified according to the most recent Oeko-Tex® Standard 100.

[c] Formaldehyde limit listed as “not detectable” is 16 parts per million.

APPENDIX IV: TOP TEN EXPORTERS OF CLOTHING TO THE UNITED STATES IN 2008

Country	Percentage of total U.S. imports[a]
China	34.3
Vietnam	6.7
Bangladesh	6.3
Honduras	5.9
Indonesia	4.8
Mexico	4.6
Cambodia	3.9
India	3.9
El Salvador	3.7
Pakistan	3.1

Source: GAO analysis of the Department of Commerce Office for Textiles and Apparel data.

Notes: Data include all apparel imports under notional category 1.

[a] By square meter equivalents.

APPENDIX V: RESULTS OF FORMALDEHYDE TESTS FOR A SAMPLE OF CLOTHING AND BED LINENS SOLD IN THE UNITED STATES

Table 5 shows the results of tests we had performed on 180 items—165 articles of clothing and 15 bed linen items.

Table 5. Information on Items Sold in the United States and Tested for Formaldehyde Levels, 2010

Comparison to most stringent regulatory standards for formaldehyde[a]	Item type	Fiber content identified on label	Fabric performance characteristic identified on label or packaging	Country identified on label	Target customer	Formaldehyde level inparts per million[b]	
						Japanese test	AATCC test
Exceeds	Dress shirt	100% cotton	Wrinkle free	China	Men	206.1	c
Exceeds	Hat	100% cotton exclusive of decoration	None identified	China	Boys 3 years of age and older	192.6	c
Exceeds	Bed linens (pillow cases)	60% cotton, 40% polyester	Soft finish, easy care	Bahrain	Adults or children 3 years of age and older	189.6	550.7
Exceeds	Khakis	100% cotton	No iron, permanent crease	India	Men	169.6	c
Exceeds	Dress shirt	60% cotton, 40% polyester	None identified	China	Boys 3 years of age and older	95.1	c
Exceeds	Bed linens (pillow cases)	100% cotton	Wrinkle free, easy care, no ironing needed, eco-friendly	USA of fabric imported from Pakistan	Adults or children 3 years of age and older	93.8	c
Exceeds	Dress shirt	100% cotton	Noniron	Indonesia	Men	92.6	c
Exceeds	Bed linens (pillow cases)	100% cotton	Wrinkle free performance	Pakistan	Adults or children 3 years of age and older	89.3	c

Comparison to most stringent regulatory standards for formaldehyde[a]	Item type	Fiber content identified on label	Fabric performance characteristic identified on label or packaging	Country identified on label	Target customer	Formaldehyde level inparts per million[b]	
						Japanese test	AATCC test
Exceeds	Bed linens (crib sheet)	100% cotton	Preshrunk	Thailand	Infants/ toddlersd	85.4	c
Exceeds	U.S. military combat uniform pants	50% cotton, 50% nylon	None identified	USA	Women	75.4	205.5
Meets	Bed linens (pillow cases)	100% cotton	None identified	China	Adults or children 3 years of age and older	72.4	c
Meets	U.S. military combat uniform pants	50% cotton, 50% nylon	Insect repellent	USA	Men	71.0	c
Meets	Dress shirt	100% cotton	Noniron, stain resistant	Indonesia	Men	63.0	c
Meets	U.S. military combat uniform shirt	50% cotton, 50% nylon	None identified	USA	Women	57.8	c
Meets	Dress pants	100% cotton	No iron, permanent crease	Vietnam	Men	54.3	c
Meets	Hat	79% cotton, 19% polyester, 2% other	None identified	China	Women	50.1	c
Meets	Bed linens (pillow cases)	100% cotton	Wrinkle free, bleach friendly	India	Adults or children 3 years of age and older	47.9	c
Meets	Khakis	100% cotton	Wrinkle free, easy care	Dominican Republic	Boys 3 years of age and older	45.9	c

Table 5. (Continued)

Comparison to most stringent regulatory standards for formaldehyde[a]	Item type	Fiber content identified on label	Fabric performance characteristic identified on label or packaging	Country identified on label	Target customer	Formaldehyde level inparts per million[b]	
						Japanese test	AATCC test
Meets	Polo shirt	100% cotton	Minimal shrinkage, fade and pill resistant, nonroll collar	Pakistan	Men	45.5	c
Meets	T-shirt	100% cotton	None identified	USA	Women	42.5	32.7
Meets	Dress pants	100% worsted wool (polyester lining)	Wrinkle and stain resistant	Mexico	Men	41.7	c
Meets	Dress shirt	100% cotton	Noniron, stain resistant	Honduras	Men	38.0	c
Meets	Dress shirt	100% cotton	Easy iron	Bangladesh	Men	37.2	c
Meets	Bed linens (sham)	60% cotton, 40% polyester	None identified	Pakistan	Boys 3 years of age and older	35.9	c
Meets	Dress shirt	100% cotton	Wrinkle resist	Bangladesh	Men	32.7	c
Meets	Jeans	98% cotton, 2% spandex	None identified	China	Women	24.0	c
Meets	Khakis	100% cotton	Stain resistant, wrinkle free	Assembled in Nicaragua	Men	23.7	43.2
Meets	Blouse	69% cotton, 27% nylon, 4% spandex	None identified	China	Women	23.5	c
Meets	Dress shirt	60% cotton, 40% polyester	Easy care, wrinkle resist	Bangladesh	Men	22.5	33.1
Meets	Hat	100% cotton	None identified	China	Women	22.5	c

Comparison to most stringent regulatory standards for formaldehyde[a]	Item type	Fiber content identified on label	Fabric performance characteristic identified on label or packaging	Country identified on label	Target customer	Formaldehyde level inparts per million[b]	
						Japanese test	AATCC test
Meets	Khakis	100% cotton	No iron, wrinkle free, easy care	Bangladesh	Men	22.4	c
Meets	Scout skirt	100% cotton	None identified	USA	Girls 3 years of age and older	21.6	c
Meets	Sweatpants	100% cotton	None identified	Honduras	Men	21.0	c
Meets	Khakis	97% cotton, 3% elastane	None identified	India	Women	20.5	c
Meets	U.S. military dress uniform shirt	65% polyester, 35% cotton	None identified	USA	Women	20.3	c
Meets	Pajamas	100% cotton	None identified	Thailand	Infants/ toddlers[d]	Not detectable	24.1
Meets	Dress shirt	60% cotton, 40% polyester	Wrinkle free, stain repellent	Vietnam	Men	Not detectable	21.8
Meets	Dress pants	90% cotton, 10% cashmere	None identified	USA of imported fabric	Boys 3 years of age and older	Not detectable	21.1
Meets	Underwear	100% cotton	None identified	India	Women	Not detectable	14.0
Meets	Bed linens (crib sheet)	100% cotton sateen	None identified	China	Infants/ toddlers[d]	Not detectable	13.8
Meets	Dress pants	100% cotton	Shrink resistant, wrinkle resistant, stain repellant, soil release	Vietnam	Men	Not detectable	11.1

Table 5. (Continued)

Comparison to most stringent regulatory standards for formaldehyde[a]	Item type	Fiber content identified on label	Fabric performance characteristic identified on label or packaging	Country identified on label	Target customer	Formaldehyde level inparts per million[b]	
						Japanese test	AATCC test
Meets	Dress	54% cotton, 45% polyester, 1% metallic. Lining: 65% polyester, 35% cotton	None identified	Bangladesh	Girls 3 years of age and older	Not detectable	11.0
Meets	Bed linens (crib sheet)	100% cotton	Easy care	China	Infants/ toddlers[d]	Not detectable	c
Meets	Cargo pants	100% cotton	None identified	Kenya	Infants/ toddlers[d]	Not detectable	c
Meets	Cargo pants	100% cotton	None identified	Indonesia	Infants/ toddlers[d]	Not detectable	c
Meets	Cloth diapers	100% cotton	Extra absorbent and quick drying	China	Infants/ toddlers[d]	Not detectable	c
Meets	Cloth diapers	100% cotton	Extra absorbent and quick drying	China	Infants/ toddlers[d]	Not detectable	c
Meets	Hat	100% cotton	None identified	China	Infants/ toddlers[d]	Not detectable	c
Meets	Hat	100% cotton exclusive of decoration	None identified	China	Infants/ toddlers[d]	Not detectable	c
Meets	Jeans	100% cotton exclusive of decoration	None identified	Indonesia	Infants/ toddlers[d]	Not detectable	c
Meets	Jeans	99% cotton, 1% spandex	None identified	Bangladesh	Infants/ toddlers[d]	Not detectable	c

Comparison to most stringent regulatory standards for formaldehyde[a]	Item type	Fiber content identified on label	Fabric performance characteristic identified on label or packaging	Country identified on label	Target customer	Formaldehyde level inparts per million[b]	
						Japanese test	AATCC test
Meets	Pajamas	100% cotton	None identified	China	Infants/ toddlers[d]	Not detectable	c
Meets	Pajamas	100% cotton	None identified	China	Infants/ toddlers[d]	Not detectable	c
Meets	Pajamas	100% cotton exclusive of decoration	None identified	China	Infants/ toddlers[d]	Not detectable	c
Meets	Romper	100% combed cotton	None identified	USA	Infants/ toddlers[d]	Not detectable	c
Meets	Romper	100% cotton	None identified	China	Infants/ toddlers[d]	Not detectable	c
Meets	Romper	100% cotton	None identified	China	Infants/ toddlers[d]	Not detectable	c
Meets	Romper	100% cotton	None identified	China	Infants/ toddlers[d]	Not detectable	c
Meets	Set (dress and leggings)	100% cotton	None identified	China	Infants/ toddlers[d]	Not detectable	Not detectable
Meets	Set (romper, pants, and bib)	100% cotton/80% cotton, 20% polyester	None identified	China	Infants/ toddlers[d]	Not detectable	c
Meets	Set (top and bottom)	100% cotton	None identified	China	Infants/ toddlers[d]	Not detectable	c
Meets	Set (top and bottom)	100% cotton	None identified	Thailand	Infants/ toddlers[d]	Not detectable	c
Meets	Set (top and bottom)	Shell: 100% cotton exclusive of decoration	None identified	China	Infants/ toddlers[d]	Not detectable	c

Table 5. (Continued)

Comparison to most stringent regulatory standards for formaldehyde[a]	Item type	Fiber content identified on label	Fabric performance characteristic identified on label or packaging	Country identified on label	Target customer	Formaldehyde level inparts per million[b]	
						Japanese test	AATCC test
Meets	Set (tops and bottom)	100% cotton	None identified	China	Infants/ toddlers[d]	Not detectable	c
Meets	Socks	82% cotton, 15% nylon, 2% spandex, 1% rubber	None identified	China	Infants/ toddlers[d]	Not detectable	c
Meets	Sweatpants	100% cotton	None identified	Bangladesh	Infants/ toddlers[d]	Not detectable	c
Meets	Sweatpants	100% cotton	None identified	China	Infants/ toddlers[d]	Not detectable	c
Meets	T-shirt	100% cotton	None identified	China	Infants/ toddlers[d]	Not detectable	c
Meets	T-shirt	100% cotton	None identified	El Salvador	Infants/ toddlers[d]	Not detectable	c
Meets	T-shirt	100% cotton	None identified	Guatemala	Infants/ toddlers[d]	Not detectable	c
Meets	T-shirt	100% cotton	None identified	Guatemala	Infants/ toddlers[d]	Not detectable	c
Meets	T-shirt	100% cotton	None identified	Thailand	Infants/ toddlers[d]	Not detectable	c
Meets	Bed linens (pillow cases)	60% cotton, 40% polyester	None identified	Pakistan	Boys 3 years of age and older	Not detectable	c
Meets	Cargo pants	100% cotton	None identified	Bangladesh	Boys 3 years of age and older	Not detectable	c

Comparison to most stringent regulatory standards for formaldehyde[a]	Item type	Fiber content identified on label	Fabric performance characteristic identified on label or packaging	Country identified on label	Target customer	Formaldehyde level inparts per million[b]	
						Japanese test	AATCC test
Meets	Dress pants	55% linen, 45% rayon	None identified	China	Boys 3 years of age and older	Not detectable	c
Meets	Jeans	100% cotton	None identified	Bangladesh	Boys 3 years of age and older	Not detectable	c
Meets	Pajamas	100% cotton	None identified	China	Boys 3 years of age and older	Not detectable	c
Meets	Polo shirt	100% organic cotton	None identified	India	Boys 3 years of age and older	Not detectable	c
Meets	Polo shirt	60% cotton, 40% polyester	None identified	Lesotho	Boys 3 years of age and older	Not detectable	c
Meets	Scout pants	67% cotton, 33% polyester	None identified	USA	Boys 3 years of age and older	Not detectable	c
Meets	Scout shirt	67% cotton, 33% polyester	None identified	Bangladesh	Boys 3 years of age and older	Not detectable	c
Meets	Sweatpants (shorts)	100% cotton	None identified	Vietnam	Boys 3 years of age and older	Not detectable	c
Meets	T-shirt	100% cotton	None identified	Honduras	Boys 3 years of age and older	Not detectable	c

Table 5. (Continued)

Comparison to most stringent regulatory standards for formaldehyde[a]	Item type	Fiber content identified on label	Fabric performance characteristic identified on label or packaging	Country identified on label	Target customer	Formaldehyde level inparts per million[b]	
						Japanese test	AATCC test
Meets	T-shirt	100% cotton	None identified	Honduras	Boys 3 years of age and older	Not detectable	c
Meets	Underwear	100% cotton	None identified	China	Boys 3 years of age and older	Not detectable	Notdetectable
Meets	Bed linens (pillow cases)	60% cotton, 40% polyester	None identified	Pakistan	Girls 3 years of age and older	Not detectable	c
Meets	Blouse	100% cotton	None identified	China	Girls 3 years of age and older	Not detectable	c
Meets	Bra	95% cotton, 5% spandex	None identified	Bangladesh	Girls 3 years of age and older	Not detectable	c
Meets	Bra	95% cotton, 5% spandex, inner cup: 100% polyester (padded)	None identified	China	Girls 3 years of age and older	Not detectable	c
Meets	Dress	100% cotton	None identified	Vietnam	Girls 3 years of age and older	Not detectable	c

Comparison to most stringent regulatory standards for formaldehyde[a]	Item type	Fiber content identified on label	Fabric performance characteristic identified on label or packaging	Country identified on label	Target customer	Formaldehyde level inparts per million[b]	
						Japanese test	AATCC test
Meets	Dress	100% cotton. Mesh—100% polyester, exclusive of decoration	None identified	China	Girls 3 years of age and older	Not detectable	c
Meets	Dress pants	97% cotton, 3% spandex	None identified	China	Girls 3 years of age and older	Not detectable	c
Meets	Jeans	56% ramie, 25% cotton, 18% polyester, 1% spandex	None identified	China	Girls 3 years of age and older	Not detectable	Not detectable
Meets	Khakis	100% cotton	None identified	Bangladesh	Girls 3 years of age and older	Not detectable	c
Meets	Polo shirt	100% cotton exclusive of decoration	None identified	Vietnam	Girls 3 years of age and older	Not detectable	c
Meets	Scout shirt	65% polyester, 35% cotton	None identified	USA	Girls 3 years of age and older	Not detectable	c
Meets	T-shirt	100% cotton	None identified	Mexico of USA Parts	Girls 3 years of age and older	Not detectable	c
Meets	T-shirt	95% cotton, 5% spandex	None identified	USA	Girls 3 years of age and older	Not detectable	c

Table 5. (Continued)

Comparison to most stringent regulatory standards for formaldehyde[a]	Item type	Fiber content identified on label	Fabric performance characteristic identified on label or packaging	Country identified on label	Target customer	Formaldehyde level inparts per million[b]	
						Japanese test	AATCC test
Meets	Underwear	95% cotton, 5% spandex	None identified	China	Girls 3 years of age and older	Not detectable	c
Meets	Bed linens (fitted sheet)	100% cotton exclusive of decoration	None identified	China	Adults or children 3 years of age and older	Not detectable	c
Meets	Bed linens (flat sheet)	60% pima cotton, 40% polyester	Easy care	China	Adults or children 3 years of age and older	Not detectable	c
Meets	Bed linens (pillow cases)	100% cotton	None identified	China	Adults or children 3 years of age and older	Not detectable	c
Meets	Bed linens (pillow cases)	60% cotton, 40% polyester	Easy care, wrinkle resistant	China	Adults or children 3 years of age and older	Not detectable	c
Meets	Cargo pants	100% cotton	None identified	Egypt	Men	Not detectable	c
Meets	Dress pants	100% wool	None identified	Vietnam	Men	Not detectable	c
Meets	Dress pants	60% cotton, 40% polyester	Easy care	Bangladesh	Men	Not detectable	c

Comparison to most stringent regulatory standards for formaldehyde[a]	Item type	Fiber content identified on label	Fabric performance characteristic identified on label or packaging	Country identified on label	Target customer	Formaldehyde level inparts per million[b]	
						Japanese test	AATCC test
Meets	Dress pants	Shell: 100% cotton	None identified	China	Men	Not detectable	c
Meets	Dress shirt	100% cotton	Easy iron	Bangladesh	Men	Not detectable	c
Meets	Dress shirt	100% cotton	Easy care	Bangladesh	Men	Not detectable	c
Meets	Dress shirt	100% cotton	None identified	USA	Men	Not detectable	c
Meets	Dress shirt	55% cotton, 45% polyester	Wrinkle free	Vietnam	Men	Not detectable	c
Meets	Dress shirt	60% cotton, 40% polyester exclusive of decoration	Easy care	Bangladesh	Men	Not detectable	c
Meets	Hat	100% cotton	Moisture management	China	Men	Not detectable	c
Meets	Hat	100% cotton exclusive of decoration	None identified	China	Men	Not detectable	c
Meets	Hat	97% cotton, 3% P.U. spandex	None identified	Bangladesh	Men	Not detectable	Not detectable
Meets	Jeans	100% Cotton	None identified	India	Men	Not detectable	c
Meets	Jeans	100% cotton	None identified	Mexico	Men	Not detectable	c
Meets	Jeans	100% cotton	None identified	Mexico	Men	Not detectable	c

Table 5. (Continued)

Comparison to most stringent regulatory standards for formaldehyde[a]	Item type	Fiber content identified on label	Fabric performance characteristic identified on label or packaging	Country identified on label	Target customer	Formaldehyde level inparts per million[b]	
						Japanese test	AATCC test
Meets	Khakis	100% combed cotton	Wrinkle resistant	Bangladesh	Men	Not detectable	c
Meets	Khakis	100% combed cotton	None identified	USA	Men	Not detectable	c
Meets	Khakis	100% cotton	No iron, wrinkle free, permanent crease, polished finish	Vietnam	Men	Not detectable	c
Meets	Khakis	100% cotton	Shrink resistant, wrinkle resistant, stain repellant, soil release	Vietnam	Men	Not detectable	c
Meets	Pajama pants	100% cotton	None identified	Cambodia	Men	Not detectable	c
Meets	Polo shirt	100% cotton	None identified	India	Men	Not detectable	c
Meets	Polo shirt	100% cotton exclusive of decoration	None identified	India	Men	Not detectable	c
Meets	Polo shirt	60% cotton, 40% polyester	Easy care	Indonesia	Men	Not detectable	c
Meets	Polo shirt	65% cotton, 35% polyester	Easy care, wrinkle resistant, super soft touch	China	Men	Not detectable	c

Meets	Socks	84% cotton, 15% nylon, 1% spandex	None identified	USA	Men	Not detectable	c
Meets	Sweatpants (shorts)	100% cotton	None identified	Honduras	Men	Not detectable	c
Meets	T-shirt	100% cotton	None identified	China	Men	Not detectable	c
Meets	T-shirt	100% cotton	None identified	Mexico	Men	Not detectable	c
Meets	T-shirt	100% cotton	None identified	Pakistan	Men	Not detectable	c
Meets	T-shirt	100% cotton	None identified	USA	Men	Not detectable	c
Meets	U.S. military combat uniform shirt	50% cotton, 50% nylon	Insect repellent	USA	Men	Not detectable	c
Meets	U.S. military dress uniform shirt	75% polyester, 25% wool	None identified	USA	Men	Not detectable	c
Meets	Underwear	100% combed cotton	Antimicrobial, pre-shrunk	Bangladesh	Men	Not detectable	c
Meets	Underwear	100% cotton	None identified	Bangladesh	Men	Not detectable	c
Meets	Underwear	100% cotton	None identified	Honduras	Men	Not detectable	c
Meets	Blouse	100% cotton	None identified	Indonesia	Women	Not detectable	Not detectable
Meets	Blouse	65% cotton, 35% silk	None identified	China	Women	Not detectable	c
Meets	Blouse	68% cotton, 28% nylon, 4% spandex	None identified	China	Women	Not detectable	c

Table 5. (Continued)

Comparison to most stringent regulatory standards for formaldehyde[a]	Item type	Fiber content identified on label	Fabric performance characteristic identified on label or packaging	Country identified on label	Target customer	Formaldehyde level inparts per million[b]	
						Japanese test	AATCC test
Meets	Bra	95% cotton, 5% lycra	None identified	Bangladesh	Women	Not detectable	c
Meets	Bra	Body/cup: 80% nylon, 20% lycra. Back lining: 84% polyester, 16% lycra	None identified	China	Women	Not detectable	Not detectable
Meets	Cargo pants	97% cotton, 3% spandex	None identified	China	Women	Not detectable	c
Meets	Cargo pants	98% cotton, 2% spandex	None identified	Bangladesh	Women	Not detectable	c
Meets	Dress	100% cotton	None identified	China	Women	Not detectable	c
Meets	Dress	100% cotton	None identified	Indonesia	Women	Not detectable	c
Meets	Dress pants	54% linen, 44% cotton, 2% spandex	None identified	Vietnam	Women	Not detectable	c
Meets	Dress pants	97% cotton, 3% spandex	None identified	Vietnam	Women	Not detectable	c
Meets	Dress pants	97% wool, 3% spandex	None identified	China	Women	Not detectable	c
Meets	Dress pants	98% cotton, 2% spandex	None identified	China	Women	Not detectable	c
Meets	Dress shirt	100% cotton	None identified	Bangladesh	Women	Not detectable	c

Comparison to most stringent regulatory standards for formaldehyde[a]	Item type	Fiber content identified on label	Fabric performance characteristic identified on label or packaging	Country identified on label	Target customer	Formaldehyde level inparts per million[b]	
						Japanese test	AATCC test
Meets	Dress shirt	100% cotton	None identified	Sri Lanka	Women	Not detectable	c
Meets	Dress shirt	100% cotton	None identified	USA	Women	Not detectable	c
Meets	Dress shirt	60% cotton, 37% polyester, 3% spandex	Easy care	China	Women	Not detectable	c
Meets	Dress shirt	96% cotton, 4% spandex	None identified	Bangladesh	Women	Not detectable	c
Meets	Dress shirt	96% cotton, 4% spandex	None identified	China	Women	Not detectable	Not detectable
Meets	Dress shirt	97% cotton, 3% spandex	None identified	China	Women	Not detectable	c
Meets	Dress shirt	97% cotton, 3% spandex	None identified	China	Women	Not detectable	c
Meets	Dress shirt	97% cotton, 3% spandex	None identified	Indonesia	Women	Not detectable	c
Meets	Dress skirt	71% polyester, 26% rayon, 3% spandex	Washable stretch, minimum care required	China	Women	Not detectable	c
Meets	Dress skirt	97% cotton, 3% spandex	None identified	Sri Lanka	Women	Not detectable	Not detectable
Meets	Jeans	100% cotton	None identified	USA	Women	Not detectable	c
Meets	Khakis	96% cotton, 4% spandex	None identified	Indonesia	Women	Not detectable	c

Table 5. (Continued).

Comparison to most stringent regulatory standards for formaldehyde[a]	Item type	Fiber content identified on label	Fabric performance characteristic identified on label or packaging	Country identified on label	Target customer	Formaldehyde level inparts per million[b]	
						Japanese test	AATCC test
Meets	Khakis	97% cotton, 3% spandex	None identified	Vietnam	Women	Not detectable	c
Meets	Khakis	97% cotton, 3% spandex	None identified	China	Women	Not detectable	c
Meets	Khakis	98% cotton, 2% spandex	None identified	China	Women	Not detectable	c
Meets	Khakis	98% cotton, 2% spandex	None identified	Vietnam	Women	Not detectable	c
Meets	Polo shirt	100% organic cotton	Soft hand, shrink resistant	Vietnam	Women	Not detectable	c
Meets	Polo shirt	95% pima cotton, 5% spandex	None identified	Vietnam	Women	Not detectable	c
Meets	Socks	76% cotton, 21% nylon, 3% lycra	None identified	Pakistan	Women	Not detectable	c
Meets	Socks	77% cotton, 18% polyester, 3% natural latex rubber, 1% spandex	None identified	USA	Women	Not detectable	c
Meets	Sweatpants	85% cotton, 15% polyester	None identified	Cambodia	Women	Not detectable	c
Meets	Sweatpants	90% cotton, 10% spandex	None identified	Bangladesh	Women	Not detectable	c
Meets	T-shirt	100% cotton	None identified	Assembled in Guatemala	Women	Not detectable	c

Comparison to most stringent regulatory standards for formaldehyde[a]	Item type	Fiber content identified on label	Fabric performance characteristic identified on label or packaging	Country identified on label	Target customer	Formaldehyde level inparts per million[b]	
						Japanese test	AATCC test
Meets	T-shirt	100% cotton	None identified	China	Women	Not detectable	Not detectable
Meets	T-shirt	100% cotton	Pre-shrunk	Honduras	Women	Not detectable	c
Meets	T-shirt	100% cotton	None identified	Thailand	Women	Not detectable	c
Meets	Underwear	92% cotton, 8% spandex	None identified	China	Women	Not detectable	c
Meets	Underwear	95% cotton, 5% spandex	None identified	Bangladesh	Women	Not detectable	c

Source: GAO analysis of information provided on items' labels or packaging and test data from an accredited commercial laboratory.

[a]As compared with the most stringent regulatory standards for formaldehyde identified by the American Apparel and Footwear Association, *Restricted Substance List*, Release 6, 2010.

[b]Formaldehyde levels listed as "not detectable" are below 20 parts per million for the Japanese test and 10 parts per million for the AATCC test.

[c]This item was tested only by the Japanese test.

[d]Infants/toddlers refers to children younger than 3 years of age.

End Notes

[1] In this report, the term "resin" encompasses both the older resin technologies that may release high levels of formaldehyde and the more recent cross-linking agents that may release little to no formaldehyde.

[2] Formaldehyde is one of five substances CPSC has identified as a strong sensitizer—a substance that can cause hypersensitivity through recurring contact—under the Federal Hazardous Substances Act. Therefore, formaldehyde and any products containing 1 percent of formaldehyde or more (10,000 parts per million) are required to bear a warning label. The 1 percent refers to the concentration of formaldehyde in solutions and products, not in the air. This reporting level for formaldehyde far exceeds amounts likely to be found in clothing.

[3] Pub. L. No. 110-314, § 234 (2008).

[4] Most of the relevant articles identified from our literature review were from medical journals or addressed medical issues, and in this report, we use the term "medical literature" to refer to the articles.

[5] American Apparel and Footwear Association, *Restricted Substance List*, Release 6 (2010).

[6] A nonprobability sample is a sample in which some items in the population have no chance, or an unknown chance, of being selected. Results from nonprobability samples cannot be used to make inferences about a population.

[7] *Medical Management Guidelines for Formaldehyde* (Atlanta: U.S. Department of Health and Human Services, Agency for Toxic Substances and Disease Registry, Feb. 7, 2008), http://www.atsdr.cdc.gov/mhmi/mmg111.html (accessed Aug. 5, 2009).

[8] U.S. Department of Health and Human Services, Public Health Service, National Toxicology Program, *Report on Carcinogens, Background Document for Formaldehyde* (Research Triangle Park, North Carolina, 2010).

[9] In this report, we use the terms "toxic" and "highly toxic" as they are used by EPA's Toxics Release Inventory Program and HHS's Agency for Toxic Substances and Disease Registry, respectively.

[10] EPA initiated an advanced notice of a proposed rulemaking in December 2008 indicating that the agency intends to investigate whether and what type of federal regulation or other action might be appropriate to protect against the risks posed by formaldehyde emitted from composite and pressed-wood products. Furthermore, on July 7, 2010, the Formaldehyde Standards for Composite Wood Act was signed into federal law. The new law amends the Toxic Substances Control Act to establish standards for emissions of formaldehyde from hardwood plywood, medium-density fiberboard, and particleboard. EPA must promulgate regulations to implement the standards by January 2013.

[11] In this report, "durable press" is used to encompass terms such as wrinkle resistant, wrinkle free, noniron, no iron, and easy care. The durable press fabric characteristic applies to items treated to retain their shape and pressed appearance after many uses, washing, and tumble drying. All-synthetic fabrics, such as 100 percent polyester, are inherently durable press and do not need to be treated.

[12] Formaldehyde may also be used in binders for prints, in various coatings such as fire retardant chemicals, and for other purposes.

[13] The International Organization for Standardization (ISO) has equivalent tests: ISO 14184-1 Textiles—Determination of Formaldehyde—Part 1: Free and Hydrolyzed Formaldehyde (Water Extraction Method), which is equivalent to the Japanese test; and ISO 14184-2 Textiles—Determination of Formaldehyde—Part 2: Released Formaldehyde (Vapour Absorption Method), which is equivalent to the AATCC test.

[14] Formaldehyde Release from Fabric, Determination of: Sealed Jar Method, AATCC Test Method 112-2008.

[15] In general, the AATCC test results in higher formaldehyde levels than the Japanese test—except at low formaldehyde levels, such as 30 parts per million or less—although the difference between the formaldehyde levels measured by the two tests depends, in part, on the type of fabric and resin used.

[16] In the *Restricted Substance List*, the American Apparel and Footwear Association, a national trade association, provides information about the regulatory limits for various substances in countries in which its members may operate. The *Restricted Substance List* identifies Austria, China, Finland, France, Germany, Japan, Lithuania, the Netherlands, New Zealand, Norway, Poland, Russia, and South Korea as countries that regulate formaldehyde in apparel, home textiles, and footwear products.

[17] Information on countries that require disclosure was compiled by the Hong Kong Standards and Testing Centre, a laboratory that provides formaldehyde testing services according to international standards and regulations.

[18] Various international agreements (or quotas) capping textile and apparel imports into the United States were put in place, in part, to attempt to curb the trend toward offshore production, which many U.S. manufacturers were using in order to control costs. The World Trade Organization agreement was put in place to lift these quotas.

[19] This organization is now called the Hamner Institutes for Health Sciences.

[20] The hematopoietic and lymphatic systems are, among other things, involved in the production of blood and in the immune system function.

[21] U.S. Department of Health and Human Services, Public Health Service, National Toxicology Program, *Report on Carcinogens, 11th Edition* (Research Triangle Park, North Carolina, 2005).

[22] 74 *Fed. Reg.* 67883 (Dec. 21, 2009).

[23] EPA, Office of Pesticides and Toxic Substances, *Assessment of Health Risks to Garment Workers and Certain Home Residents From Exposure to Formaldehyde* (Washington, D.C., 1987); and *Integrated Risk Information System, Formaldehyde (CASRN 50-00-0)* (Washington, D.C.: EPA), http://www.epa.gov/ncea/iris/subst/0419.htm (downloaded Dec. 9, 2009).

[24] EPA, *IRIS Toxicological Review of Formaldehyde—Inhalation Assessment (External Review Draft)* (Washington, D.C., June 2010).

[25] The National Academies comprises four organizations: the National Academy of Sciences, the National Academy of Engineering, the Institute of Medicine, and the National Research Council.

[26] World Health Organization, International Agency for Research on Cancer, *IARC Monographs on the Evaluation of Carcinogenic Risks to Humans, Volume 88, Formaldehyde, 2-Butoxyethanol and 1-tert-Butoxypropan-2-ol* (Lyon, France, 2006).

[27] Robert Baan et al. "A Review of Human Carcinogens—Part F: Chemical Agents and Related Occupations," *The Lancet Oncology*, vol. 10, issue 12 (2009).

[28] Department of Energy, Oak Ridge National Laboratory, *Formaldehyde Release From Durable-Press Apparel Textiles, Final Project Report to the U.S. Consumer Product Safety Commission* (Oak Ridge, Tennessee, 1985).

[29] Dermal exposure occurs when skin comes in contact with free formaldehyde in textiles. Formaldehyde in clothing may be either bound or free, and it is the skin contact with free formaldehyde that is associated with the negative health effects. Free formaldehyde occurs when there is an incomplete binding process between the formaldehyde-based resin and the

clothing fibers or when the resin decomposes. Unless otherwise noted in this report, references to the health effects associated with formaldehyde in clothing or textiles refer to free formaldehyde.

[30] Anton C. De Groot and Howard I. Maibach, "Does Allergic Contact Dermatitis from Formaldehyde in Clothes Treated with Durable-Press Chemical Finishes Exist in the USA?" *Contact Dermatitis*, vol. 62, no. 3 (2009).

[31] Ryan M. Carlson, Mary C. Smith, and Susan T. Nedorost, "Diagnosis and Treatment of Dermatitis Due to Formaldehyde Resins in Clothing," *Dermatitis*, vol. 15, no. 4 (2004).

[32] Etain Cronin, *Contact Dermatitis* (New York City: Churchill Livingstone, 1980).

[33] While a national regulation of formaldehyde in clothing would not necessarily produce comprehensive data on levels in clothing, such data are unlikely to be compiled without a requirement to do so.

[34] We considered all items in our sample to come into direct contact with the skin, although some textile industry standards consider bed linens for older children and adults to be subject to the formaldehyde limits applicable to items that do not come into direct contact with the skin.

[35] We are using "durable press" to include terms such as wrinkle resistant, wrinkle free, noniron, no iron, and easy care.

[36] European Commission Directorate General Joint Research Centre, Physical and Chemical Exposure Unit, Chemical Release from Textiles, *European Survey on the Release of Formaldehyde from Textiles* (Ispra, Italy, 2007).

[37] *Evaluation of Alleged Unacceptable Formaldehyde Levels in Clothing* (Wellington, New Zealand: New Zealand Ministry of Consumer Affairs, Oct. 17, 2007), http://www.consumeraffairs.govt.nz/legislation-policy/policy-reports-and-papers/reports (accessed Oct. 22, 2009). The New Zealand government also received information from four retailers that tested 203 items. One item, a preproduction fabric, was found to be above the acceptable level, which was defined by New Zealand to be 100 parts per million.

[38] *No Formaldehyde Found in Clothing,* Australian Competition and Consumer Commission, (October 17, 2007), http://www.accc.gov.au/content/index.phtml?itemId=801314 (accessed Aug. 5, 2009).

[39] B.A. Kottes Andrews, "Wrinkle Resistant Cotton and Formaldehyde Release," *Colourage Annual* (1995).

[40] *Formaldehyde Release from Durable-Press Apparel Textiles, Final Project Report to the U.S. Consumer Product Safety Commission* (1985).

[41] The current OSHA regulation, among other things, limits airborne exposure to 0.75 parts of formaldehyde per million parts of air over an 8-hour workday; sets a short-term exposure limit of 2 parts of formaldehyde per million parts of air; and includes a hazard communication requirement pertaining to formaldehyde gas, all mixtures or solutions composed of greater than 0.1 percent formaldehyde (equivalent to 1,000 parts per million), and materials capable of releasing formaldehyde into the air, under reasonably foreseeable conditions of use, at concentrations reaching or exceeding 0.1 parts per million. The National Institute for Occupational Safety and Health has established nonenforceable guidelines for formaldehyde exposure and recommends an occupational exposure limit over an 8-hour workday of 0.016 parts per million and a 15-minute ceiling of 0.1 parts per million in the air.

[42] This formaldehyde level in textiles is based on the Japanese test.

[43] Eero Priha, “Are Textile Formaldehyde Regulations Reasonable? Experiences from the Finnish Textile and Clothing Industries,” *Regulatory Toxicology and Pharmacology*, 22, 243-249 (1995).

[44] Department of Health and Human Services, National Institute for Occupational Safety and Health, *The Registry of Toxic Effects of Chemical Substances, No. LP8925000, CAS No. 50-00-0* (Washington, D.C., 2009).

[45] The countries the American Apparel and Footwear Association identifies as having regulatory standards for formaldehyde in clothing and other textiles are Austria, China, Finland, France, Germany, Japan, Lithuania, the Netherlands, New Zealand, Norway, Poland, Russia, and South Korea.

[46] The Austrian Textile Research Institute and the German Research Institute Hohenstein jointly developed the Oeko-Tex® Standard 100 in 1992. The International Oeko-Tex® Association, which includes 14 textile research and test institutes in Europe and Japan, is responsible for the independent tests for harmful substances according to Oeko-Tex® Standard 100.

[47] We requested information on corporate limits on formaldehyde in clothing and other textiles, if any, from 16 publicly traded U.S.-based clothing retailers. One of the 14 retailers responding reported that it had not established formaldehyde limits. For one of the two retailers that did not respond, we were able to obtain, from its Web site, information about the corporate limits it has established.

[48] According to medical literature, moderate amounts of formaldehyde are defined as between 1,000 and 100 parts per million, while low and ultra-low levels are below 100 parts per million, based on the AATCC test.

End Notes for Appendix II

[1] Pub. L. No. 110-314, § 234 (2008).

[2] U.S. Department of Health and Human Services, National Toxicology Program, *Report on Carcinogens, 11th Edition* (Research Triangle Park, North Carolina, 2005).

[3] U.S. Department of Health and Human Services, Public Health Service, Agency for Toxic Substances and Disease Registry, *Toxicological Profile for Formaldehyde* (Atlanta, Georgia, 1999).

[4] *Integrated Risk Information System, Formaldehyde (CASRN 50-00-0)* (Washington, D.C.: EPA), http://www.epa.gov/ncea/iris/subst/0419.htm (downloaded Dec. 9, 2009).

[5] U.S. Environmental Protection Agency, *External Review Draft: Toxicological Review of Formaldehyde—Inhalation Assessment, in Support of Summary Information on the Integrated Risk Information System, Volume I of IV* (Washington, D.C., June 2010).

[6] World Health Organization, International Agency for Research on Cancer, *IARC Monographs on the Evaluation of Carcinogenic Risks to Humans, Volume 88, Formaldehyde, 2-Butoxyethanol and 1-tert-Butoxypropan-2-ol* (Lyon, France, 2006).

[7] Organisation for Economic Co-operation and Development, *SIDS Initial Assessment Report for 14th SIAM, Formaldehyde (CAS No. 50-00-0)* (Paris, France, 2002).

[8] A nonprobability sample is a sample in which some items in the population have no chance, or an unknown chance, of being selected. Results from nonprobability samples cannot be used to make inferences about a population.

[9] The American Association for Laboratory Accreditation bases its accreditation on the International Organization for Standardization (ISO) 17025:2005, *General requirements for the competence of testing and calibration laboratories*.

[10] American Apparel and Footwear Association, *Restricted Substance List*, Release 6 (2010).

End Notes for Appendix III

[1] We have identified a few U.S.-based retailers that comply with some voluntary programs, but they do not generally sell individual items that are labeled to indicate compliance.

In: Thread Threat: Formaldehyde in Textiles ISBN: 978-1-61324-839-3
Editor: Victoria C. Muñoz

Chapter 2

TESTIMONY OF RUTH A. ETZEL, MD, PHD, FAAP, ON BEHALF OF THE AMERICAN ACADEMY OF PEDIATRICS, BEFORE THE SUBCOMMITTEE ON CONSUMER PROTECTION, PRODUCT SAFETY, AND INSURANCE, HEARING ON "FORMALDEHYDE IN TEXTILES AND CONSUMER PRODUCTS"*

Good morning. I appreciate this opportunity to testify today before the Commerce, Science and Transportation Subcommittee on Consumer Protection, Product Safety and Insurance regarding formaldehyde in textiles and consumer products. My name is Ruth Etzel, MD, PhD, FAAP, and I am proud to represent the American Academy of Pediatrics (AAP), a non-profit professional organization of more than 60,000 primary care pediatricians, pediatric medical sub-specialists, and pediatric surgical specialists dedicated to the health, safety, and well-being of infants, children, adolescents, and young adults. I am the Founding Editor of the AAP's book on Pediatric Environmental Health, and I am currently editing a 3rd edition. I am also a

* This is an edited, reformatted and augmented version of testimony given by Ruth A. Etzel on behalf of the American Academy of Pediatrics, before the Subcommittee on Consumer Protection, Product Safety, and Insurance, Hearing on "Formaldehyde in Textiles and Consumer Products", on April 28, 2009.

former Chair of the AAP Committee on Environmental Health and the founding chair of the AAP Section on Epidemiology.

Formaldehyde is a toxic, pungent, water-soluble gas used in the aqueous form as a disinfectant, fixative, or tissue preservative, making it versatile for a wide range of uses. Formaldehyde resins are used in wood products (e.g. particleboard, paper towels), plastics, paints, manmade fibers (e.g. carpets, polyester), cosmetics, and other consumer products,[1] including many with which children have regular contact.[2] According to recent research and media reports, formaldehyde may be found in fabrics and children's clothing[3], children's furniture,[4] baby bath products,[5] and other products. Formaldehyde is also used in the resins used to bond laminated wood products and to bind wood chips in particleboard. Particleboard may be used in various types of furniture, including cribs and other items meant for use by or with children. The experience of Gulf Coast families living in mobile homes and travel trailers after Hurricane Katrina brought these hazards to the nation's attention; trailers, which have small, enclosed spaces, low air exchange rates, and many particleboard furnishings, may have much higher concentrations of formaldehyde than other types of homes.[6,7]

Formaldehyde gas is known to cause a wide range of health effects. A common air pollutant in the home,[8] formaldehyde is an eye, skin, and respiratory tract irritant. In other words, it can cause burning or tingling sensations in the eyes, nose and throat. Children may be more susceptible than adults to the respiratory effects of formaldehyde. Even at fairly low concentrations, formaldehyde can produce rapid onset of nose and throat irritation, causing cough, chest pain, shortness of breath, and wheezing. At higher levels of exposure, it can cause significant inflammation of the lower respiratory tract, which may result in swelling of the throat, inflammation of the windpipe and bronchi, narrowing of the bronchi, inflammation of the lungs, and accumulation of fluid in the lungs. Pulmonary injury may continue to worsen for 12 hours or more after exposure. Children may be more vulnerable than adults to the effects of chemicals like formaldehyde because of the relatively smaller diameter of their airways. Children may be more vulnerable because they breathe more rapidly than adults for their size, and they may be developmentally incapable of evacuating an area promptly when exposed.[9]

Formaldehyde may exacerbate asthma in some infants and children. Studies since 1990 have found higher rates of asthma, chronic bronchitis, and allergies in children exposed to elevated levels of formaldehyde.[10,11,12,13]

In 2004, the International Agency for Research on Cancer (IARC) announced there was sufficient evidence that formaldehyde causes nasopharyngeal cancer in humans and reclassified it as a Group 1, known human carcinogen (previous classification: Group 2A). IARC also reported there was limited evidence that formaldehyde exposure causes nasal cavity and paranasal cavity cancer and "strong but not sufficient" evidence linking formaldehyde exposure to leukemia.[14] The U.S. National Toxicology Program classifies it as "reasonably anticipated to be a human carcinogen."[15]

Formaldehyde can cause contact dermatitis in susceptible people. Dr. Brookstein will discuss this matter in more detail, so I will only note that children are as susceptible as adults to the dermal effects of formaldehyde exposure.

Due to its toxicity, various nations have taken steps to limit the use of formaldehyde in some applications. Several nations have set standards for the presence of formaldehyde residues in fabric, including Finland, Norway, the Netherlands, and Germany. The European Union limits formaldehyde in children's clothing to 30 parts per million.[16] Other nations, such as Japan, China, Russia, Lithuania, New Zealand, and South Korea have set limits on formaldehyde in textiles and/or wood products. Among these nations, the strongest restrictions are in place in Japan, which requires no detectable residue of formaldehyde in clothing for children birth to 3 years of age.[17]

RECOMMENDATIONS

The American Academy of Pediatrics has made formaldehyde recommendations to Congress and the Administration in the past, and would like to reiterate those and submit others for Congress's consideration.

CPSC Should Limit Formaldehyde Residues in Children's Clothing and Other Products

Given that at least a dozen other nations already restrict formaldehyde residues in children's clothing, CPSC should collaborate with EPA and other agencies with scientific and medical expertise to determine similar limits to be imposed in the U.S. While more research is needed to refine our understanding of formaldehyde's impact on child health, there is already a considerable body of evidence that may be sufficient to allow CPSC to make a reasonable judgment in this area. The agency should also require labels on children's clothing and products that indicate the presence of formaldehyde residues.

More Research is Needed on Formaldehyde and Children's Health

In July 2007, the Academy suggested to the House of Representatives Committee on Energy and Commerce that the Federal Emergency Management Agency and federal health agencies undertake a systematic, scientifically rigorous study of this issue to determine children's exposure levels and correlation with reported symptoms, and steps that should be taken to safeguard their health. To our knowledge, no such study has been conceived or implemented. It also remains unclear to what extent children may be exposed to formaldehyde from multiple sources, and what effect this may have on their developing bodies. The Consumer Product Safety Improvement Act of 2008 requires the Consumer Product Safety Commission (CPSC) Comptroller General to conduct a study within two years of "the use of formaldehyde in the manufacture of textile and apparel articles...to identify any risks to consumers caused by the use of formaldehyde in the manufacturing of such articles..." This report is due in January 2011.

EPA Should Adopt Nationwide California's Proposed Restrictions on Formaldehyde Emissions from Wood Products

In January 2009, the AAP joined numerous other organizations in urging Environmental Protection Agency Administrator Lisa Jackson to adopt nationwide the restrictions on formaldehyde emissions from hardwood plywood, particleboard, and medium density fiberboard set under the California Air Resource Board Airborne Toxics Control Measure.

CPSC Should Develop Educational Materials for Consumers About Formaldehyde and its Presence and Role in Various Products, as Well as Potential Health Risks

The CPSC could provide an important service by providing up-to-date educational materials about formaldehyde. A search of the agency's website reveals a number of documents about formaldehyde, but many of them are from the 1970s and 1980s. The last version of the most comprehensive document, "An Update on Formaldehyde," appears to be the 1997 revision.[18]

The American Academy of Pediatrics commends you, Mr. Chairman, for holding this hearing today to call attention to the potential hazards of formaldehyde exposure among children. We look forward to working with Congress to minimize the exposure of children and all Americans to all potentially toxic chemicals. I appreciate this opportunity to testify, and I will be pleased to answer any questions you may have.

End Notes

[1] International Agency for Research on Cancer. IARC Monographs on the Evaluation of Carcinogenic Risks to Humans. Volume 88. Formaldehyde. Available online at http://monographs.iarc.fr/ENG/Monographs/vol88/volume88.pdf.

[2] Kelly TJ, Smith DL, Satola J. Emission Rates of Formaldehyde from Materials and Consumer Products Found in California Homes. *Environ Sci Technol*, 1999;33(1): 81-88.

[3] "Poison found in kids' clothes from China." New Zealand Sunday Star-Times, August 19, 2007. Available online at http://www.stuff.co.nz/sunday-star-times/497.

[4] Environment California Research & Policy Center. Toxic Baby Furniture: The Latest Case for Making Products Safe from the Start. May 2008. Available online at http://www.environmentamerica.org/uploads/MF/Uh/MFUhMHLNuROm0SNHVkLkxg/Toxic-Baby-Furniture---The-Latest-Case-for-Making-Products-Safe-from-the-Start.pdf.

[5] Environmental Working Group. No More Toxic Tub: Getting Contaminants Out of Children's Bath and Personal Care Products. March 2009. Available online at http://www.ewg.org/node/27698.

[6] American Academy of Pediatrics Committee on Environmental Health. Air Pollutants, Indoor. In: Etzel, RA, ed. Pediatric Environmental Health, 2d Edition. Elk Grove Village: American Academy of Pediatrics, 2003.

[7] Spengler JD. Sources and concentrations of indoor air pollution. In: Samet JM, Spengler JD, eds. *Indoor Air Pollution: A Health Perspective.* Baltimore, MD: Johns Hopkins University Press; 1991.

[8] American Academy of Pediatrics Committee on Environmental Health. Air Pollutants, Indoor. In: Etzel, RA, ed. Pediatric Environmental Health, 2d Edition. Elk Grove Village: American Academy of Pediatrics, 2003.

[9] Agency for Toxic Substances & Disease Registry. Medical Management Guidelines for Formaldehyde. http://www.atsdr.cdc.gov/MHMI/mmg111.html#bookmark02

[10] American Academy of Pediatrics Committee on Environmental Health. Air Pollutants, Indoor. In: Etzel, RA, ed. Pediatric Environmental Health, 2d Edition. Elk Grove Village: American Academy of Pediatrics, 2003.

[11] Wantke F, Demmer CM, Tappler P, Gotz M, Jarisch R. Exposure to gaseous formaldehyde induces IgE-mediated sensitization to formaldehyde in school-children. *Clin Exp Allergy.* 1996 Mar; 26(3):276-80.

[12] Garrett MH, Hooper MA, Hooper BM, Rayment PR, Abramson MJ. Increased risk of allergy in children due to formaldehyde exposure in homes. *Allergy*. 1999 Apr; 54(4):330-7.

[13] Rumchev KB, Spickett JT, Bulsara MK, Phillips MR, Stick SM. Domestic exposure to formaldehyde significantly increases the risk of asthma in young children. *Eur Respir J.* 2002 Aug; 20(2):403-8.

[14] International Agency for Research on Cancer, "IARC Classifies Formaldehyde As Carcinogenic to Humans," Press Release No. 153, June 15, 2004, http://www.iarc.fr/ENG/Press Releases/archives/pr153a.html

[15] Krzyzanowski M, Quackenboss JJ, Lebowitz MD. Chronic respiratory effects of indoor formaldehyde exposure. *Environ Res.* 1990 Aug;52(2):117-25.

[16] Information on European Union laws regarding limits on formaldehyde in textiles available online from the Centre for the Promotion of Imports from developing countries, http://www.cbi.eu/.

[17] American Apparel and Footwear Association. Restricted Substances List. February 2009. Available online at http://www.apparelandfootwear.org/UserFiles/File/ Restricted%20Substance %20List/AAFARSLRele ase4Feb09.pdf.

[18] U.S. Consumer Product Safety Commission. An Update on Formaldehyde, 1997 Revision. Available online at http://www.cpsc.gov/CPSCPUB/PUBS/725.pdf.

In: Thread Threat: Formaldehyde in Textiles ISBN: 978-1-61324-839-3
Editor: Victoria C. Muñoz

Chapter 3

TESTIMONY OF DAVID BROOKSTEIN, SC.D., DEAN AND PROFESSOR, SCHOOL OF ENGINEERING & TEXTILES, PHILADELPHIA UNIVERSITY, BEFORE THE SUBCOMMITTEE ON CONSUMER PROTECTION, PRODUCT SAFETY AND INSURANCE, HEARING ON "USE OF FORMALDEHYDE AND OTHER TOXIC MATERIALS IN TEXTILES AND APPAREL"*

Thank you Chairman Pryor and members of the Committee for this opportunity to provide testimony to the Senate Subcommittee on Consumer Protection, Product Safety and Insurance. I would also like to express my appreciation to Senator Robert P. Casey, Jr. who is at the vanguard of protecting our nation's citizens from potentially toxic materials in consumer products. My testimony is based on over 35 years of experience as a textile engineering professor and researcher including co-founding the Institute for Textile and Apparel Product Safety at Philadelphia University.

* This is an edited, reformatted and augmented version of testimony given by David Brookstein, of the School of Engineering & Textiles, Philadelphia University, before the Subcommittee on Consumer Protection, Product Safety and Insurance, Hearing on "Use of Formaldehyde and other Toxic Materials in Textiles and Apparel", on April 28, 2009.

In the summer of 2007 reports surfaced about high levels of lead in toys and other consumer goods and there were hundreds of thousands of items recalled. One area that initially escaped scrutiny at that time was textile and apparel product safety. Years before, the federal government recognized the lethal toxicity of asbestos fibers and TRIS flame retardant in children's sleepwear and acted appropriately to ban their use in consumer products. Today, once again, the question of safety is front and center and researchers are looking for answers regarding the safety of textiles and apparel. By researching the prevalence of other potentially toxic chemicals, such as formaldehyde, dyes and finishes, used every day in clothing, we will be able to determine just what chemicals and at what levels could pose risks to all of us, especially our children — and possibly lead to medical conditions ranging from contact dermatitis to neurotoxicity, endocrine disruption and possibly cancer.

Many clothing items are in direct contact with the skin. During contact there can be perspiration which involves moisture transport between the skin and the dyed and chemically treated clothing items. Dyes are used to enhance the appearance of textiles and chemical treatments affect the performance of textile products. While modern dyes and chemical treatments are chemically bound to the fibers in the clothing, there is the possibility that residual dye (dye bleed) and finishes (treatment chemicals) are released in direct contact with the skin. Textile materials are a capillary and porous material with different pore sizes, and can be saturated with both liquid and gaseous water during wear. The transportation of perspiration through this material at different temperatures is a very complex process, which can involve convection, capillary flow, penetration, molecular diffusion, evaporation, and solidification.

On Aug 14, 2008 Public Law 110-314 (Consumer Product Safety Improvement Act) was enacted. The purpose of the law was to establish consumer product safety standards and other safety requirements for children's products and to reauthorize and modernize the Consumer Product Safety Commission.

Formaldehyde is a commonly used chemical treatment for apparel items and has long been recognized as toxic. Accordingly, Senators Casey, Brown, Clinton and Landrieu offered an Amendment to study the use of formaldehyde in manufacturing textile and apparel articles. The Amendment, agreed to unanimously, calls for a study by the GAO in consultation with the Commission, on the use of formaldehyde in the manufacture of textile and apparel articles, or in any component of such articles, to identify any risks to consumers caused by the use of formaldehyde in the manufacturing of such articles, or components of such articles. The law calls for the study to be completed by August 2010 but, to our knowledge, the GAO has not yet begun the study.

Formaldehyde treatment of cellulosic fibers such as cotton was first taught in an invention by the British inventors Foulds, Marsh and Wood in US Patent 1,734,516 in 1929. The inventors claimed that "one of the greatest defects of a fabric composed entirely of cotton has been the ease with which such fabric is creased or crumpled when crushed or folded under pressure in the hand." The invention was to use a mixture of chemicals including formaldehyde to cause a chemical reaction with the cellulose that would cause cross-linking and thus render the fabric wrinkle free.

Substantial commercial interest developed as inherently wrinkle-free synthetic fibers were commercialized and by the 1950's family fabric caretakers (mostly women) were delighted by the potential of wrinkle-free fabrics that would add to other labor-saving chores that were being introduced to the public. As more and more women joined the workforce the entire family became interested in easy care clothing.

In 1985, The US National Institute for Occupational Safety and Health (NIOSH) completed its first research study of formaldehyde. The study examined death certificates among 256 deceased workers from three plants which made shirts from formaldehyde treated cloth. Formaldehyde was used at these plants to help make shirts more crease resistant. The 1985 study found a significantly increased risk of cancer of the buccal cavity (cancer of the inside of the mouth) and for multiple myeloma (cancer of the bone marrow). In 1988, NIOSH completed its second study of formaldehyde exposure. This study looked at employment records from 11,030 workers who had been employed at any one of three plants. Two of the three plants were the same as in the previous study. As in the 1985 study, the 1988 study found a significantly increased risk for cancer of the buccal cavity. Excess risks were also seen for multiple myeloma and leukemia.

In 2004, NIOSH conducted a substantially large study of cause of death among clothing workers exposed to formaldehyde and found that:

1) The death rates from all causes combined and for all cancers combined among the 11,039 workers in the updated study were lower than expected, based on the U.S. population rates.

2) There were no deaths from cancers of the nasopharynx (nose). The death rate for cancer of the buccal cavity (inside of the mouth) was only slightly elevated.
3) The overall risk for myeloid leukemia was almost 11/2 times what was expected.
4) For workers who were employed at the plants for 10 or more years and were first exposed 20 years earlier, the risk for myeloid leukemia was increased over 2 times what was expected.
5) The increase in myeloid leukemia was also seen among those workers who were first exposed prior to 1963, when formaldehyde exposures were likely higher.

NIOSH reported that the overall average concentration of formaldehyde measured by NIOSH at the three plants during the early 1980's was 0.15 parts per million (ppm). This was below the permissible level at that time, which was 3.0 ppm over an 8-hour work day. Exposures were similar across departments and plants. In 1987 the permissible level of formaldehyde exposure was reduced to 1.0 ppm and in 1992 was further reduced to 0.75 ppm. OSHA regulation 29 CFR 1910-1048 regulates the exposure limit for workers in the US textile and apparel industry to 1 part formaldehyde per million parts of air as an 8-h time-weighted average. The NIOSH study was based on a group of scientific research papers published from 1985-2004.[1,2,3]

While the NIOSH studies and subsequent regulations were directed at American workers, the same concerns obtain for American consumers.

In 2004, the World Health Organization International Agency for Research on Cancer (IARC) categorized formaldehyde as a known cancer-causing agent in humans.

The United States apparel manufacturing industry has declined precipitously and today it has been estimated that approximately 90% of consumer apparel sold in the United States is not manufactured in the United States. Accordingly, today the safety hazards associated with formaldehyde to US apparel workers is negligible, if any. Yet while there are essentially no occupational hazards associated with formaldehyde processing of apparel to US workers there could be hazards to those overseas workers who produce clothing and textiles for the US marketplace. Additionally, American workers can be exposed to potential toxic off-gassing from textile products when imported items are received in US distribution centers.

However, humans can be exposed to formaldehyde associated with textiles and clothing in an additional manner than that from manufacturing.

For instance, in the clothes treated with formaldehyde can come into direct contact with the skin. In 1959, Marcussen (Denmark) reported that during a period between 1934-1958 there were 26 cases (11% of studied cases) of garment formaldehyde dermatitis.[4] Marcussen also reported results of a study conducted from1934-1955 a study in which 1-3% of 36,000 eczematous patients showed formaldehyde sensitivity.[5] In 1965, US dermatology researchers O'Quinn and Kennedy reported contact dermatitis caused by formaldehyde in clothing.[6] Hatch published a complete review of references to clothing based formaldehyde sensitivity in 1984.[7] The medical literature is replete with many studies showing the adverse dermatological effects of formaldehyde. An excellent current review of this subject has been written by Fowler "Formaldehyde as a Textile Allergen" in 2003. 8

On the next page is a table which shows common formaldehyde resins used in textiles and apparel.

Resin Type	Relative Formaldehyde Release*
Urea formaldehyde/DMU	High
Melamine formaldehyde	High
DMDHEU (Fixapret CPN)	Low
DMDHEU blended or reacted with glycols (modified) (Fixapret ECO)	Very low
Dimethoxymethyl dihydroxyethylene urea (methylated DMDHEU)	Very low
Dimethyl dihydroxyethylene urea (Fixapret NF)	None

* High signifies a formaldehyde release of > 1,000 ppm; low, a release of < 100 ppm; and very low, a release of < 30 ppm.[9]

At a recent workshop held at Philadelphia University attended by personnel from the Consumer Product Safety Commission, Dr. Susan Nederost of University Hospitals of Cleveland/Case Western Reserve University reported that patients with allergic contact dermatitis, such as that caused by allergic response to formaldehyde exposure, results in substantial amount of days missed from employment.

Another exposure route is from off-gassing of stored or closeted clothing with relatively high levels of formaldehyde. As early as 1960 researchers reported on release of formaldehyde vapors on storage of wrinkle-resistant cotton fabrics.[10] The exposure route from off-gassing of formaldehyde could soon be recognized as a significant health risk to United States consumers as a result of recent testimony to the US House of Representatives which reports the relatively high levels of formaldehyde in house and office blackout shades and other drapery items.[11] Using the AATCC Test Method #112 free formaldehyde values of between 1000 ppm and 3000 ppm were found in a relatively large group of imported items available in the United States marketplace.

As of yet, there are no formaldehyde restrictions or standards for clothing and other textile items that are distributed and sold in the United States. However more and more nations are adopting standards for formaldehyde in clothing and textiles. In Japan, textile fabrics are required by law to contain less than 75 ppm free formaldehyde, as measured by the method described in Japan Law 112. And no formaldehyde is tolerated for infant clothing. The Hong Kong Standards and Testing Center produced the table below which shows the status of formaldehyde regulations in countries that are currently addressing this situation.[12] From the table, the Committee can easily see how other industrialized countries are dealing with this important issue that affects the health of their citizenry.

In addition Poland, Russia, Lithuania and South Korea now regulate formaldehyde in textiles and apparel.

Formaldehyde is also found in glues and adhesive used to bond materials to each other such as in layers of shoes and fabrics to each other. In particular, para-tertiary butylphenol (PTBP) formaldehyde resin is sometimes used. This type of formaldehyde resin can also cause allergic reactions.[13]

Some have suggested that one way for the consumer to deal with residual formaldehyde on newly purchased clothing is to just wash it prior to wearing it. This is fundamentally problematic since many consumers will not heed this labeling "suggestion" and will just wear newly purchased clothing without taking the time to wash it. Additionally, further scientific evidence needs to be obtained that shows there is no residual formaldehyde on clothing even after its been washed. And finally, there are many items where formaldehyde is used and there is no opportunity for pre-washing. These items include baseball caps and footwear.

While currently there are no US standards or regulations associated with formaldehyde in clothing and textiles the American Apparel and Footwear Association (AAFA) published a 2008 Restricted Substance List (RSL) which was refined in 2009. AAFA requested that its members abide voluntarily to the standards listed. For formaldehyde the RSL suggests no detectable formaldehyde for infant clothing (0-36 months), 75 ppm for clothing in direct contact with skin (>36 months) and 300 ppm for textiles with no direct skin contact (>36 months).

Country	Regulations / Requirements	Objection Limit / Limit
Germany	Gefahrstoffverordnung (Hazardous Substances Ordinance) Annex III, No. 9, 26.10.1993	Textiles that normally come into contact with the skin and release more than 1500 mg/kg formaldehyde must bear the label "Contains formaldehyde. Washing this garment is recommended prior to first time use in order to avoid irritation of the skin."
France	Official Gazette of the French Republic, Notification 97/0141/F	The regulations apply to products that are intended to come into contact with human skin, including textiles, leather, shoes, etc. Textiles for babies: 20 mg/kg Textiles in direct skin contact: 100 mg/kg Textiles not in direct skin contact: 400 mg/kg
Netherlands	The Dutch (Commodities Act) Regulations on Formaldehyde in Textiles (July 2000)	Textiles in direct skin contact must be labeled "Wash before first use" if they contain more than 120 mg/kg formaldehyde and the product must not contain more than 120 mg/kg formaldehyde after wash.
Austria	Formaldehydverordnung, BGBL Nr. 194/1990	Textiles that contains 1500 mg/kg or above must be labeled.
Finland	Decree on Maximum Amounts of Formaldehyde in Certain Textiles Products (Decree 210/1988)	Textiles for babies under 2-year-old: 30 mg/kg Textiles in direct skin contact: 100 mg/kg Textiles not in direct skin contact: 300 mg/kg
Norway	Regulations Governing the Use of a Number of Chemicals in Textiles (April 1999)	Textiles for babies under 2-year-old: 30 mg/kg Textiles in direct skin contact: 100 mg/kg Textiles not in direct skin contact: 300 mg/kg
China	Limits of Formaldehyde Content in Textiles GB18401-2001	Textiles for infants and babies: ≤20 mg/kg Textiles in direct skin contact: ≤75 mg/kg Textiles not in direct skin contact: ≤300 mg/kg
Japan	Japanese Law 112	Textiles for Infants: not detectable Textiles in direct skin contact: 75 ppm

In addition to formaldehyde in textiles and apparel, there are other well documented toxic chemicals that are used in clothing, furniture and other textile-based consumer items. In particular, there are two classes of dyes that are commonly used in consumer textile-based products that are widely recognized as having the potential to cause allergic contact dermatitis and possibly to cause cancer. These two dye classes are azoic (azo) and disperse dyes. There is such a widespread concern associated with the use of azo dyes in textile-based products that many countries have enacted restrictive standards and stringent regulations that limit their use. In 2002 the European Union published a Directive (2002/61/EC) to restrict the marketing and use of certain dangerous substances and preparations (azo colorants) in textile and leather products Thus, in the European Union their use is regulated by law; in the United States, at this time, there exist only *voluntary standards* by those companies that agree to regulate their use.

In 2006 a series of previously unreported cases of dermatitis appeared in Finland. Rantanen, a Finnish physician, reported that by 2007 “many cases from all over the country” were reported in the internet discussion forum of the Finnish Dermatological Society. After an extensive investigation it was found that the cases were due to exposure to dimethylfumarate (DMF).[14] It was reported by British newspaper accounts that sachets of DMF were put in thousands of Chinese manufactured furniture items to prevent mold while in storage or while being transported.[15] Rantenen reported that the patients showed strong positive patch test reactions to upholstery fabric samples and to dimethylfumarate, down to a level of 1 ppm in the most severe case. It was concluded that the cause of the Chinese sofa/chair dermatitis epidemic was likely to be allergy to dimethylfumarate, a novel potent contact sensitizer. Thus, a serious health issue can occur, not from the furniture fabric but from the release of allergenic agents contained in the foam cushioning. As can be seen from the picture of a patient exposed to DMF the condition presents itself in a most devastating manner.

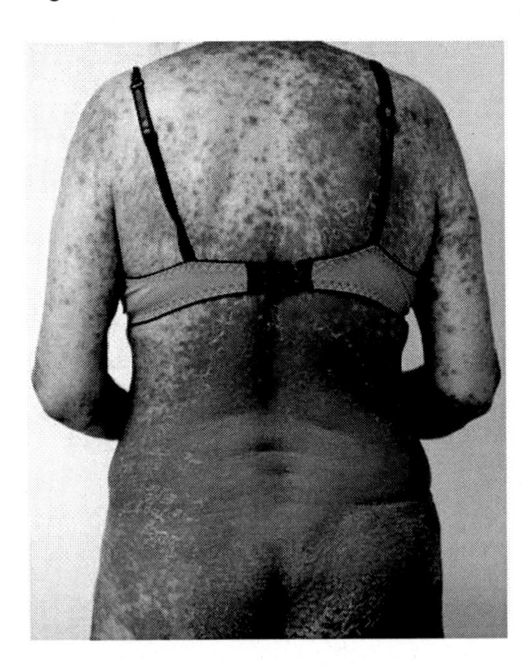

Patient Exposed to Dimethylfumurate in Sofa.

The European Union acknowledged the dangers of using dimethyl-fumurate in consumer products and issued European Directive (2009/251/EC) on March 17, 2009. The directive requires that products containing DMF are not to be placed on the market. The Directive also requires any product containing DMF that has already been placed on the market be withdrawn by May 1, 2009 and that consumers be made aware of the potential risks.

Brominated chemicals, used to make fabrics flame retardant, are another class of toxic substances that is of great concern to researchers. Of particular concern to child safety advocates are flame retardant fabrics used in children's car seats. While flame retardant fabrics play a beneficial role in preventing or minimizing serious injury, the long-term harmful effects to children exposed to this class of toxic chemicals is unknown and should be a matter for further research.

Unfortunately, a recent study conducted at Philadelphia University using an X-Ray Fluorescence analyzer showed a range of bromine readings from about 0.43% to 0.86%. It is widely recognized by the research community that levels in excess of 0.1% are considered toxic. Consequently, this standard has been adopted by the European Union in the Restriction of Hazardous Substances (RoHS) standards. The RoHS Directive is an EU Legal Directive for environmental regulations concerning the Restriction of Use of Hazardous Substances. The Directive requires the removal of five hazardous substances from electric and electronic equipment (Pb, Cd, Cr, Hg, Br compounds). While these toxic compounds are restricted in electric and electronic equipment, we were concerned that the same chemical compounds might be used in children's car seats. Accordingly, an extensive chemical analysis of the fabric was conducted to determine the bromine compounds that were present in car seat fabric with relatively high levels of bromine. Two specific brominated compounds were found: Hexabromocyclododecane (HBCD) – 0.425% and Tetrabromobisphenol A (TBBPA) – 1.185%.

HBCDs are included on the OSPAR[16] list of chemicals for priority action. HBCDs have been identified by the U.K. Chemical Stakeholders Forum as persistent, bioaccumulative and toxic.[17] While currently no specific regulatory actions are being taken in the United States, HBCDs have been identified for risk assessment in Canada Australia and Japan. Further regulatory/assessment activities in these countries will take place over the next few years.[18]

Studies suggest that HBCD affects thyroid hormone levels, causes learning and memory defects in neonatal laboratory animals, and has been detected in breast milk.[19] There are indications that oral exposure to HBCDs induces drug-metabolizing enzymes in rats, such as hepatic cytochrome P450

(CYP),[20] and that HBCDs may induce cancer by a nonmutagenic mechanism.[21,22] There are reports that HBCDs can disrupt the thyroid hormone system[23] and affect the thyroid hormone receptor-mediated gene expression.[24] Following neonatal exposure experiments in rats, developmental neurotoxic effects can be induced, such as aberrations in spontaneous behavior, learning, and memory function.[25] HBCDs can also alter the normal uptake of neurotransmitters in rat brains.[26]

TBBPAs are included on the OSPAR list of chemicals for priority action. TBBPA is known to off-gas to the environment, though the amount of off gassing varies depending how the TBBPA was combined with other materials.[27] Lab tests have suggested that it may disrupt thyroid function.[28] Studies also suggest that it may adversely affect hormone levels and the immune system.[29] Histological findings showed that the slight enlargement of the hepatocytes, inflammatory cell infiltrations and focal necrosis of hepatocytes were more marked in liver of treated groups (from 350 mg/kg Body Weight) than in control group. The present data suggest the possibility of inducing hepatic lesions by TBBPA.[30]

In view of my testimony and the wide body of knowledge associated with the use of toxic chemicals in textiles and apparel I believe that now is the time to look again at the issue of formaldehyde and other potential toxic dyes and finishes in textiles and apparel. It is recommended that future legislation dealing with consumer product safety should include a study on the use of formaldehyde and other known toxic dyes, finishes, and preservatives in the manufacture of textile and apparel articles, that consumer product safety standards be implemented based on the findings of these studies, and a reasonable testing program be established for textile and apparel items including components of such articles in which formaldehyde and other known toxic chemicals were used in their manufacture.

The suggested study of the use of toxic chemicals in textiles and apparel products will provide Congress the needed information to consider whether new laws and /or regulations are necessary to protect the health and welfare of American citizens.

In conclusion, I would like to again express my appreciation to the Committee and to Senator Casey for this opportunity to provide testimony on this important issue that affects the health of our citizenry. I stand ready to serve the Committee in any way in the future.

End Notes

[1] Stayner L, Smith AB, Reeve G, et al. Proportionate mortality study of workers in the garment industry exposed to formaldehyde. Am J Ind Med 1985;7:229-240.

[2] Stayner LT, Elliott L, Blade L, et al. A retrospective cohort mortality study of workers exposed to formaldehyde in the garment industry. Am J Ind Med 1988;13:667-681.

3 Pinkerton LE. Hein MJ, Stayner LT. Mortality among a cohort of garment workers exposed to formaldehyde: an update. Occup Environ Med 2004;61(3):193-200.

[4] Marcussen, P V, Contact Dermatitis Due to Formaldehyde in Textiles, 1934-1958, Preliminary Report, Acta Derm. Venereol. 39,348-356 (1959).

[5] Marcussen, P V, Dermatitis Caused by Formaldehyde Resins in Textile, Dermatologica, 125, 101-111 (1962)

[6] O'Quinn, S E, and Kennedy C B, Contact Dermatitis Due to Formaldehyde in Clothing Textiles, J. Am. Med/ Soc. 194, 593-596 (1965).

[7] Hatch K L, Chemicals and Textiles, Part II: Dermatological Problems Related to Finishes, Textile Research Journal, Vol. 54, No. 11, 721-732 (1984).

[8] Fowler, JF, Formaldehyde as a Textile Allergen, *Elsner P, Hatch K, Wigger-Alberti W (eds): Textiles and the Skin.Curr Probl Dermatol. Basel, Karger, 2003, vol 31, pp 156-165.*

[9] Hatch KL, Maibach HI. Textile dermatitis: an update. (I). Resins, additives and fibers. Contact Dermatitis 1995;32:319-26.

[10] Reid J D, Arceneaux, R L et al. Studies of wrinkle resistant finishes for cotton textiles (I): Release of formaldehyde vapors on storage of wrinkle resistant cotton fabrics. Am Dyest Rep 1960: 49, 490-531.

[11] Berman M, Testimony to the Ways and Means Trade Subcommittee U.S. House of Representatives, 2007

[12] http://www.stc-group.org/UserFiles/File/Newsletter/TMD/Flormaldehyde2004.pdf

[13] Geldof B Am Roesyanto I D, Van Joost T H, Clinical aspects of para-tertiary-butlyphenol formaldehyde resin (PTFR) allergy, Contact Dermatitis, 1989, 21, 312-315.

[14] The cause of the Chinese sofa/chair dermatitis epidemic is likely to be contact allergy to dimethylfumarate, a novel potent contact sensitizer T. Rantanen British Journal of Dermatology 2008 159, pp218–221.

[15] Brown D, Thousands injured by 'toxic gas from Chinese sofas, The Times, July 21, 2008 UK

[16] The 1992 OSPAR Convention is the current instrument guiding international cooperation on the protection of the marine environment of the North-East Atlantic. It combined and up-dated the 1972 Oslo Convention on dumping waste at sea and the 1974 Paris Convention on land-based sources of marine pollution.

[17] Covaci, A.;Gereke,A;Law,R.;Voorspoels,S.;Kohler,M.;Heeb,N.;Leslie,H.;Allchin,C.;Boer,J.; Hexabromocyclododecanes (HBCDs) in the Environment and Humans: A Review. Environmental Science & Technology, 2007, vol. 40, No. 12.

[18] National Chemicals Inspectorate (KEMI) *Draft of the EU Risk Assessment Report on Hexabromocyclododecane, Sundyberg*, Sweden, 2005.

[19] Birnbaum L, Staskal D. 2004. "Brominated flame retardants: cause for concern?" *Environmental Health Perspectives* Vol. 112:1.

[20] Germer, S.; Piersma, A. H.; van der Ven, L.; Kamyschnikow, A.; Fery, Y.; Schmitz, H. J.; Schrenk, D. Subacute effects of the brominated flame retardants hexabromocyclododecane and tetrabromobisphenol-A on hepatic cytochrome P450 levels in rats. *Toxicology* 2006, *218*, 229-236.

[21] Helleday, T.; Tuominen, K. L.; Bergman, A.; Jenssen, D. Brominated flame retardants induce intragenic recombination in mammalian cells. *Mutat. Res.* 1999, *439*, 137-147.

[22] Ronisz, D.; Finne, E. F.; Karlsson, H.; Forlin, L. Effects of the brominated flame retardants hexabromocyclododecane (HBCDD) and tetrabromobisphenol-A (TBBP-A)on hepatic enzymes and other biomarkers in juvenile rainbow trout and feral eelpout. *Aquat. Toxicol.* 2004, *69*, 229-245.

[23] Eriksson, P.; Viberg, H.; Fischer, C.; Wallin, M.; Fredriksson, A. A comparison on developmental neurotoxic effects of hexabromocyclododecane, 2,2 ,4,4 ,5,5 - hexabromodiphenylether (PBDE 153) and 2,2,4,4,5,5-hexachlorobiphenyl (PCB 153). *Organohalogen Compd.* 2002, *57*, 389-392.

[24] Yamada-Okabe, T.; Sakai, H.; Kashima, Y.; Yamada-Okabe, H. Modulation at a cellular level of the thyroid hormone receptormediated gene expression by 1,2,5,6,9,10-hexabromocyclododecane (HBCD), 4,4-diiodobiphenyl (DIB), and nitrofen (NIF). *Toxicol. Lett.* 2005, *155*, 127-133.

[25] Eriksson, P.; Viberg, H.; Fischer, C.; Wallin, M.; Fredriksson, A. A comparison on developmental neurotoxic effects of hexabromocyclododecane, 2,2,,4,4,5,5-hexabromodiphenylether (PBDE 153) and 2,2 ,4,4 ,5,5 -hexachlorobiphenyl (PCB 153) *Organohalogen Compd.* 2002, *57*, 389-392.

[26] Mariussen, E.; Fonnum, F. The effect of brominated flame retardants on neurotransmitter uptake into rat brain synaptosomes and vesicles. *Neurochem. Int.* 2003, *43*, 533-542.

[27] Birnbaum L, Staskal D. 2004. “Brominated flame retardants: cause for concern” *Environmental Health Perspectives* Vol. 112:1.

[28] Kitamura S, Kato T, Iida M, Jinno N, Suzuki T, Ohta S, Fujimoto N, Hanada H, Kashiwagi K, Kashiwagi A. 2005.“Anti-thyroid hormonal activity of tetrabromobisphenol A, a flame retardant, and related compounds: Affinity to the mammalian thyroid hormone receptor, and effect on tadpole metamorphosis.” *Life Sciences*. 2005 Feb 18; 76(14); 1589-601.

[29] Birnbaum L, Staskal D. 2004. “Brominated flame retardants: cause for concern?” *Environmental Health Perspectives*. Vol. 112:1.

[30] Tada, Y; Fujitani,T;Ogata, A; Kamimura, H. Flame retardant tetrabomobisphenol A induced hepatic changes in ICR male mice, *Environmental Toxicology and Pharmacology*. August 2007

In: Thread Threat: Formaldehyde in Textiles ISBN: 978-1-61324-839-3
Editor: Victoria C. Muñoz

Chapter 4

TESTIMONY OF DR. PHILLIP WAKELYN, BEFORE THE SUBCOMMITTEE ON CONSUMER PROTECTION, PRODUCT SAFETY, AND INSURANCE, HEARING ON "FORMALDEHYDE IN TEXTILES AND CONSUMER PRODUCTS"*

SUMMARY

There have been no valid safety related problems raised in the US concerning the low levels of formaldehyde on clothing and textiles. In view of all the studies over the last 30 years indicting that there is not a formaldehyde problem with US textiles products and regulations already in place concerning formaldehyde and textiles, no new regulations are necessary. Because the evidence is so strong that formaldehyde in textiles does not pose a problem to consumers, there is no need for legislative or regulatory action concerning formaldehyde and textiles unless the results of the GAO study, required by Section 234 of the CPSIA which became law August 14, 2008, indicate that action is necessary.

---

* This is an edited, reformatted and augmented version of testimony given by Dr. Phillip Wakelyn, before the Subcommittee on Consumer Protection, Product Safety, and Insurance, Hearing on "Formaldehyde in Textiles and Consumer Products" on April 28, 2009, at 10:30 a.m.

1. Introduction

Allergic contact dermatitis caused by textiles is rare. There are many reasons other than chemical additives used in processing of textiles that can cause irriation/allegic contact dermatitis – the fabric itself, physical effects of the clothing rubbing the skin, heat retention from perspiration soaked clothes, poor hygiene, fasteners, and other devices attached to clothing, etc. For example, some people may find that fabrics such as wool irritate their skin but it is not an allergy and not chemically related. It is important to note that formaldehyde is ubiquitous and is a natural product present in the air from many sources -- natural processes, in fruits, vegetables and blood, by combustion processes, including motor vehicles, cooking, household heating and brush fires and produced by cigarette smoking.

2. Fabric Levels of Formaldehyde Should Not Be Confused with Airborne Levels of Formaldehyde Gas

Fabric levels of formaldehyde are determined by two generally accepted methods (see Appendix 3). Typically, fabric levels are expressed as micrograms of formaldehyde per gram of fabric (µg/g or ppm). Airborne levels are expressed as micrograms or milligram of formaldehyde gas per cubic meter of air (µg or mg/m^3; ppb or ppm). There is not a clear correlation between fabric levels of formaldehyde and airborne levels of formaldehyde gas because release mechanisms are numerous and complex. Many factors affect releases and airborne levels, e.g., material and treatment, temperature, humidity, room size, air exchanges in the room, etc. Chamber studies of textiles indicate that a 300-500 µg/g fabric level would have air emissions less than the California Proposition "safe harbor" level of 40 µg/day per textile.

The health risk of high fabric levels is dermatitis; high airborne levels can cause respiratory health problems. The CPSC in the 1980's considered urea formaldehyde foam insulation (UFFI) to be a hazardous product and took actions under the FHSA against its use. The CPSC Report, "An Update on Formaldehyde, 1997 Revision" indicates: p.3 "… Formaldehyde is one of several gases present indoors that may cause illnesses. Many of these gases, as well as colds and flu, cause similar symptoms." To reduce levels of formaldehyde from pressed wood products, mandatory formaldehyde standards for emissions from pressed wood products have been promulgated and proposed [CA Air Resources Board an airborne toxic control measure (ATCM) to reduce formaldehyde emissions from composite wood products and from finished goods that contain composite wood products (17 CA Code of Regulations, sections 93120-93120.12) passed 4/07 effective 1/1/09; US EPA, ANPR, "Formaldehyde Emissions from Pressed Wood Products", 73 FR 73620, 12/3/08].

In the 1980's CPSC determined that no standard was needed for fabric levels or textile product emissions of formaldehyde for textiles and apparel. CPSC extensively studied formaldehyde and textiles in the 1980's at the Oak Ridge National Laboratory, Research Triangle Institute, and elsewhere (see data below). After numerous studies, it was concluded that formaldehyde levels in textiles and formaldehyde emissions from textiles were so low that they do not pose an acute or chronic health hazard for consumers, i.e. that clothing/apparel does not present an unreasonable risk to consumers from formaldehyde.

According to chamber tests and other studies on a wide range of textiles/apparel products before and after washing that had been treated with formaldehyde containing chemicals/adducts, the air emissions levels of formaldehyde gas from textiles and apparel were below the level of concern. *Further, it was concluded that formaldehyde emissions* from textiles and apparel do not require a warning label under California Proposition 65 or by EPA, because test data have shown that their emissions are below the level of concern (<40 μg/day per textile).

3. Dyeing and Finishing of Textiles – Where Formaldehyde Containing Chemicals/Adducts Are Used

Textile fibers can be natural or manufactured. Natural fibers are cellulose vegetable fibers (bast, leaf, seed hairs) such as cotton or linen or protein animal fiber such as wool or silk. Manufactured fiber such as rayon and acetate are cellulose polymers; synthetic polymer fibers include nylon, polyester, polypropylene, and spandex.

Textiles go through many processes to produce a dyed and finished commercial textile. As many as twenty or more finishing treatment can be used (see WD Schindler and PJ Hauser, 2004.***Chemical finishing of textiles***,

Woodhead Publishing, Ltd). Some textile finishing processes use formaldehyde containing chemicals/adducts – for easy-care / durable press / wrinkle resistance for sheeting, shirting, dress goods, knits, and slacks; for textile pigment dyeing for a small number of sheets and for pigment printing; and for flame retardance for very little if any children's sleepwear and protective work clothing.

Formaldehyde containing chemicals/adducts are used mainly on cotton and cotton blends and other cellulosic fabrics/textiles (see Appendix 5). Easy care/wrinkle resist cotton apparel accounts for 2% of the total apparel offerings at retail and for 13% of total cotton apparel purchased in 2008. The majority easy care cotton apparel is men's apparel. There is almost no easy care children's apparel and almost no children's wear is treated with formaldehyde containing chemicals/adducts of any kind.

Formaldehyde containing chemicals/adduct finishes are not used on synthetic textiles such as fabrics/apparel/clothing made from nylon and polyester.

4. Formaldehyde and Textiles

Formaldehyde-releasing finishes provide crease resistance, dimensional stability, and flame retardance for textiles and can serve as binders in textile pigment printing and dyeing (Priha, 1995). Easy-care / durable press / wrinkle resistance finishing is one of the many finishing operations used to give finished textiles the quality and aesthetics that consumers demand. These finishes are generally applied to cellulose and cellulose blend fabrics -- fabrics used for sheeting, shirting, dress goods, knits, and slacks. The primary effects of these finishes on cellulosic fibers are reduction in swelling and shrinkage, improved wet and dry wrinkle recovery, smoothness of appearance after drying and retention of intentional creases and pleats. Commercially available apparel is not treated with formaldehyde directly to produce easy-care/ durable press / wrinkle resistant textiles. Formaldehyde has not been shown to be a useful reagent to produce wrinkle resistant cotton (Priha, 1995). Methylolamide agents (N-methylol compounds, formaldehyde adducts of amides or amide-like nitrogenous compounds), which introduce ether cross-links between cellulose molecules of the cotton fiber, are the most widely used to produce wrinkle resistant cotton [see P.J. Wakelyn, N.R. Bertoniere, A. D. French, et al. 2007. *Cotton Fiber Chemistry and Technology*. Series:

International Fiber Science and Technology, CRC Press (Taylor and Francis Group), pp. 75-76].

Durable-press/wrinkle resistant resins or permanent-press resins containing small amounts of formaldehyde have been used on cotton and cotton/polyester blend fabrics since the mid-1920s to impart wrinkle resistance during wear and laundering. Priha (1995) indicated that formaldehyde-based resins, such as urea-formaldehyde (UF) resin, were once more commonly used for crease resistance treatment. However, better finishing agents with lower formaldehyde release have been developed and are what is currently used. Totally formaldehyde-free crosslinking agents are now available but they are expensive and do not perform as well (e.g., can affect some dye shades).

There are a small amount of sheets where acrylic and acrylic- based binders that can contain traces of formaldehyde are used for pigment printing and dyeing. Very little if any halogen phosphorus flame retardants that contain formaldehyde are used on children's sleepwear and protective work clothing.

Some apparel that is treated with formaldehyde containing chemicals/adducts can potentially release trace amounts of formaldehyde, even though they are bonded to the fiber. If apparel, cotton and cotton blends and other cellulosic fabrics/textiles, are treated with formaldehyde-derived chemicals (i.e., formaldehyde adducts of amides or amide-like nitrogenous compounds, acrylic binders or halogen phosphorus flame retardant compounds), the potential trace amount of formaldehyde that could be released should be far below levels that would cause irritation or any health effects or affect the environment.

It has been reported that the average formaldehyde level contained by textiles made in the USA is approximately 100–200 μg free formaldehyde/g as measured by the AATCC Method 112 sealed jar test (results using AATCC Method 112 are about 4 times higher than that measured using ISO 14184-1/ Japanese Law 112 Method) (Scheman et al., 1998). Modern innovations through the use of derivates and scavengers and other low-emitting resin technology (Wakelyn et al. 2007 cited above) keep the levels below 100-200 ppm (as measured by AATCC 112 Method). The AATCC 112 method has been the most common way for determining formaldehyde levels in fabrics in the US but since textiles are international products ISO 14184-1 and the Japanese Law 112 Method are now being used more often.

Tests in New Zealand on Chinese textiles (see Appendix 4), which were conducted after incorrect stories reported high fabric formaldehyde levels, showed that "97 of 99 items had no detectable or very low levels of formaldehyde." "Two items had above the acceptable level of 100 parts per million, but simple washing reduced formaldehyde to well below acceptable levels."

It is easy to neutralize the formaldehyde with Clorox 2. It has been known for a long time that simple laundering with normal commercial detergents greatly reduces any formaldehyde or lowers to non-detectable levels.

Published scientific studies indicate that it is very rare for even highly sensitized individuals to have a reaction to formaldehyde fabric concentrations as low as 300 ppm [by AATCC Method 112] (Hatch and Maibach, 1995). And patch testing with formaldehyde, textile resins that can release formaldehyde, and formaldehyde-releasing preservatives lend support to the idea that the causal agent of allergic contact dermatitis due to wearing durable press fabrics may be the resin rather than formaldehyde that may be released.

- *Clothing Dermatitis and Clothing-Related Skin Conditions*, August 2001, http://www.lni.wa.gov/Safety/Research/Dermatitis/files/clothing.pdf).
- Hatch KL, Maibach HI (1995) Textile dermatitis: an update (I). Resins, additives and fibers. *Contact dermatitis*, 32:319–326.
- Priha E (1995) Are textile formaldehyde regulations reasonable? Experiences from the Finnish textile and clothing industries. *Regulatory toxicology and pharmacology*, 22:243–249.
- Scheman AJ, Carrol PA, Brown KH, Osburn AH (1998) Formaldehyde-related textile allergy: an update. *Contact dermatitis*, 38:332–336.

5. US Government Studies Regarding Formaldehyde and Textiles

Both the US Consumer Product Safety Commission (CPSC) and the US Environmental Protection Agency (EPA) have determined that no standard for fabric levels or product emissions is necessary for textiles and apparel.

CPSC extensively studied formaldehyde and textiles in the 1980's at the Oak Ridge National Laboratory, Research Triangle Institute, and elsewhere. After these studies, it was determined that formaldehyde fabric levels and formaldehyde emissions from textiles do not pose an acute or chronic health problem to consumers.

- Robins, J.D. and Norred, W.P., Bioavailability in Rabbits of Formaldehyde from Durable Press Textiles, Final Report on CPSC IAG 80-1397, USDA Toxicology and Biological Constituents Research Unit, Athens, GA, 1984.
- ORNL/TM-9790 'Formaldehyde Release from Durable-Press Apparel Textiles' Final Project Report to CPSC Oct 1985 [TG Mathews, CR Daffron, ER Merchant] http://www.ornl.gov/info/reports/1985/3445600564985.pdf
- RTI 'Percutaneous Penetration of Formaldehyde' (July 1981-83) submitted in Jan 1984 to ATMI and FI by A R Jeffcoat, RTI [rhesus monkey study] [Any formaldehyde that was release did not show up in any organs of the animal. Dr Peter Pruess previously with CPSC and now with EPA was involved these studies.}
- CPSC Briefing Package on formaldehyde and textiles "Status Report on the Formaldehyde in Textiles Portion of Dyes and Finishes Project" [Sandra Eberle (to Peter Pruess and others), 1/3/84] p. 4 Conclusions: 'current evidence, although not conclusive, does not indicate that formaldehyde exposure from resin-treated textiles is likely to present a carcinogenic hazard.'

Formaldehyde emissions from textiles do not require a warning label under CA Proposition 65.

Much work was done by the textile and cotton industries when Prop 65 was first being implemented in 1986. The textile and cotton industries resolved this issue with the CA Health and Welfare Agency in 1987 to 1992. Chamber and other studies were done with various textile products before and after washing. The state of CA indicated in a letter to the textile industry in 1988 that the state has no information that suggests that textiles pose a risk (Letter to W.A. Shaw, Textile Industry Coalition from Dr. S.A. Book, Science Advisor to the Secretary, California Health and Welfare Agency, Mar 22, 1988). The regulation of Proposition 65 is now under Office of Environmental Health Hazard Assessment (OEHHA), CA EPA. The concern in CA lately has been with emissions from wood products not textiles. As far as I am aware there has not been a bounty hunter suit in CA against apparel. No product has a "general exemption" but a product is not required to have warning labels and has no requirements under Prop 65 unless that product causes potential exposure above the “safe harbor limit” to any substance that is on the Prop 65 list. The key point is that the trace emissions of formaldehyde from an individual textile does not exceed the "safe harbor level" of 40μg/day for formaldehyde (gas) [http://oehha.ca.gov/prop65/pdf/2009FebruaryStat.pdf].

6. Conclusion

In view of all the studies over the last 30 years indicting that there is not a problem with US textiles and regulations already in place concerning formaldehyde and textiles, no new regulations are necessary. There should be no action concerning formaldehyde and textiles unless the results of the GAO study required by the CPSIA clearly show that areas of concern still exist.

Appendix 1: Formaldehyde Containing Chemicals Used In Textile and Apparel Dyeing and Finishing Are Regulated by US CPSC and Other US Regulatory Agencies

- CPSC has the authority to regulate formaldehyde under the Federal Hazardous Substances Act (15 U.S. Code 1261-1278). CPSC already has authority to regulate substances/chemicals or mixtures of substances on textiles that may cause substantial personal injury or illness during any customary or reasonably foreseeable handling or use and has a regulation [under "strong sensitzer" in section 2(k) of the Act, 16 CFR 1500.13(d) (repeated in 1500.3(b)(9))]. CPSC has banned chemicals in the past under the FHSA and investigated formaldehyde, flame retardants, dyes, and other chemicals used in preparation, dyeing, and finishing of textiles.
- EPA under the Toxic Substances Control Act (TSCA) has authority over all chemicals in commerce and can set restrictions or ban chemicals. They currently have a significant new use rule that covers any flame retardants as well as any textile chemicals. EPA also can regulate emission levels from products but is not concerned with formaldehyde emissions from textiles and apparel.
- OSHA has the authority to regulate exposures of formaldehyde within a workplace (29 CFR 1910.1048). The OSHA workplace level is 0.75 ppm (8 hr TWA). Also products containing > 0.1 % formaldehyde and "materials capable of releasing formaldehyde into the air, under

foreseeable conditions of use at concentrations reaching or exceeding 0.1 ppm are subject to regulation including labeling, worker training and MSDS's.

- California Proposition 65 [the Safe Drinking Water and Toxic Enforcement Act of 1986] requires labeling for chemicals known to the state of California to be carcinogens or reproductive toxins that cause exposures of significant risk. Product emissions of formaldehyde gas from textiles and apparel do not require labeling under California Proposition 65, because tests have shown that their emissions are below the level of concern, i.e., the "safe harbor level" for formaldehyde that does require labeling is <40 μg/day per textile. 40 μg/day per textile is negligible compared to natural background levels.
- There are also national and international voluntary standards (e.g., American Association of Textile Chemists and Colorists [AATCC], the American Society for Testing and Materials [ASTM], and International Organization Standards [ISO]) that are used in the textile industry. In addition, the American Apparel & Footwear Association [AAFA] publishes a Restricted Substances List (RSL) that many companies are using in addition to their own RSLs.
- There are also eco-labeling standards, e.g., the EU Ecolabel for Textiles, Oeko-Tex Standard 100 and sustainability standards (e,g,. NSF-336) for textiles are being developed by the American National Standards Institute (ANSI).

APPENDIX 2: INTERNATIONAL STANDARDS, COMPANY REQUIREMENTS, VOLUNTARY LABELS

There are governmental restrictions, company requirements (e.g., Levi Strauss, Marks and Spencer) and several labels (e.g., EU Ecolabel, Oeko-Tex Standard 100) that set limits for free or easily freed formaldehyde in textiles. The European eco-label for textiles [EU (2002), Ecolabel for Textiles, http://eur-lex.europa.eu/LexUriServ/LexUriServ.do?uri=OJ:L:2002:133:0029:0041:EN:PDF] has a limit of 300 ppm formaldehyde (by ISO-14186-1/Japanese Law 112 Method). Finished fabrics for adult clothing and other skin contact textiles may be labeled and called low formaldehyde finished according to Oeko-Tex Stanandard 100 when their free formaldehyde content is lower than 75 ppm (Japan Law 112 Method).

Eight counties in the world have formaldehyde requirements for textiles ranging from 1500 ppm (in Germany) to 75 ppm (in Japan measured by the Japanese Law 112 Method) for textiles that contact the skin. The other countries are 100-120 ppm (measured by the Japanese Law 112 Method/ ISO 14184-1).Discussion in the "Proposed Government Product Safety Policy Statement on Acceptable Limits of Formaldehyde in Clothing and other Textiles" by the New Zealand government [http://www.consumeraffairs.govt.nz/policylawresearch/product-safety-law/proposed-statement/proposed-policy-statement.pdf] gives a summary of International formaldehyde limits is clothing and other textiles (p. 3). International regulatory limits show a diverse spread. Japan has the most stringent limits for clothing in direct contact with the skin, 75 ppm. The section on Test Method on p. 5 first paragraph states: Below 20 ppm the result is reported as "not detectable". This is for the proposed acceptable testing method, ISO 14184-1, which is essentially the same as Japan Law 112 Method. Öko-Tex 100 defines measured values <20 ppm on the substrate according to Japan Law 112 Method as non detectable. In the AATCC Method 112 the margin of error or the "zero" level in low-level samples is 75 ppm.

APPENDIX 3: MEASURING THE AMOUNT OF FORMALDEHYDE IN TEXTILES

There are currently two generally accepted methods of measuring formaldehyde in textiles. The method used needs to be specified. It is important an acceptable testing method be used. It is the only way that meaningful data can be obtained.

- AATCC Method 112 ("sealed jar test") – Free and releasable/hydrolysable formaldehyde may be captured by this procedure. The test specimen is suspended over an aqueous solution in a sealed jar at a given temperature for a specific time. Formaldehyde gas given off is absorbed in to the aqueous solution; formaldehyde in the solution is derivatized and the color of the resulting complex is measured with a visible spectrophotometer. Formaldehyde amount is expressed as micrograms of formaldehyde per gram of fabric (μg/g or ppm). The margin for error or the "zero" level in low-level samples is about 75 ppm. This has been the predominant method used by the US Textile Industry.

- o AATCC Technical Manual, Test Method 112

- ISO-14184-1 and Japanese Law 112 Method [The ISO and the Japanese methods are essentially the same and give the same results] – Free formaldehyde is measured and probably only a small amount of releasable/hydolizable formaldehyde is measured. The formaldehyde is extracted from the specimen into water, the formaldehyde is derivatized and measured with a visible spectrophotometer as above. The limit of detection for both methods or "zero" level is 20 μg/g or ppm. The ISO Standards for testing formaldehyde provide internationally agreed methods of testing.
  - o ISO 14184-1:1998 Textiles -- Determination of formaldehyde -- Part 1: Free and hydrolized formaldehyde (water extraction method)
  - o ISO 14184-2:1998 Textiles -- Determination of formaldehyde -- Part 2: Released formaldehyde (vapour absorption method)
  - o Law for the Control of Household Products Containing Harmful Substances (Japanese Law 112) and Japanese Industrial standard (JIS) L 1041
- An AATCC Method 112 reading of 300 ppm (meeting most US retailer requirements) may give a ISO-14184-1/Japanese Method 112 value of 75 ppm – an exact correlation between the two methods is not possible. Other methods for measuring formaldehyde on fabrics have described but how they correlate with the ISO-14184-1/Japanese Law 112 Method or the AATCC 112 Method is not published.

APPENDIX 4: NEW ZEALAND TESTING IN 2007 ON CHINESE CLOTHES

http://times.busytrade.com/489/1/Chinese_Clothes_Gain_Good_Comment_From_New_Zealand.html

Chinese Clothes Gain Good Comment From New Zealand

(October 23, 2007) From:fiber2fashion

Chinese clothes gained good comment from New Zealand for its high safety index, which has much to do with the Chinese government' s Longtime effort on improving product quality. On October 17, the New Zealand Ministry of Consumer Affairs posted on its website the result of the formaldehyde test it conducted on 99 items of Chinese clothes.

According to the Ministry, among the 99 items, 97 did not contain or contained formaldehyde lower than the country's standard, and the two items that contained formaldehyde higher than the standard could lower its formaldehyde content through simple cleaning. The test result of New Zealand authority showed that Chinese clothes were safe.

We noticed the wide publicity of high formaldehyde content in Chinese clothes on New Zealand media since August this year. The test that New Zealand government conducted and the result it released proved that Chinese products were safe. China appreciated the objective attitude of New Zealand in handling this issue.

Chinese government attached great importance to product quality and safety. A series of recent measures to tighten quality control and food safety control would significantly improve the quality and reputation of Chinese products.

According to the China Customs, China exported about 290 million US dollars worth of clothes to New Zealand, accounting for 70.5% of its apparel market. In the formaldehyde test that New Zealand conducted this time, Chinese exports made up 84% of the tested clothes.

Ministry of Commerce of the People's Republic of China (MOFCOM)… http://www.fibre2fashion.com/news/textile-news/newsdetails.aspx?news_id =42744

New Zealand : Formaldehyde tests show no health issue in clothes
October 18, 2007

Test results released show little cause for concern about levels of formaldehyde in clothing and textiles on sale in New Zealand.

"In response to concerns raised by television programme Target, the Ministry of Consumer Affairs tested 99 items of clothing and manchester," says Consumer Affairs Minister Judith Tizard.

"97 of 99 items had no detectable or very low levels of formaldehyde." "Two items had above the acceptable level of 100 parts per million, but simple washing reduced formaldehyde to well below acceptable levels."

Twenty parts per million is accepted internationally as the zero mark under which formaldehyde in fabric is not detectable.

Ms Tizard says the Ministry used the correct method of testing and its results were robust and credible. “Target used the wrong testing method, which is why their results were so dramatically different.”

“In line with international best practice for testing clothing, the Ministry tested for free formaldehyde only. Target tested for combined free and bound formaldehyde. They then compared this with international standards for free formaldehyde.”

“It was like testing apples and oranges against a standard for apples only.”

The government is to issue a product safety policy statement setting acceptable levels of formaldehyde in clothing, a move that will provide greater certainty for New Zealand consumers.

“We are consulting on the appropriate levels, but expect they will be similar to those used as benchmarks in the Ministry’s testing, which were based on levels used by overseas regulators.”

Submissions on the proposed policy statement are due by 26 November.

The Ministry of Consumer Affairs have been working closely with the Australian Competition and Consumer Commission, who are today also announcing a consistent approach to acceptable levels of formaldehyde in clothing.

New Zealand Ministry of Consumer Affairs

APPENDIX 5: EASY CARE MARKET INFORMATION 1) WHAT IS OFFERED AT RETAIL, 2) WHAT THE CONSUMER IS BUYING

1) Retail Offerings

Apparel

- Easy care cotton apparel accounts for 2% of the total apparel offerings at retail.
- The majority (97%) of easy care cotton apparel is men’s apparel.

Easy Care Apparel Categories

Category	Share of Products with Easy Care
Total Men's Apparel	4%
Men's Dress Shirts	9%
Men's Casual Pants	14%
Men's Other Pants	15%

Home Textiles

- Easy care cotton apparel accounts for 1% of the total home textile offerings at retail.

Category	Share of Products with Easy Care
Bedding	1%
Sheeting	3%

Source: Cotton Incorporated's Retail Monitor™ is a quarterly survey of apparel products at 26 major US retailers. Information is collected in the store and online. In first quarter 2009, data were collected from 42,564 apparel products. The home textiles data is from the 2009 Home Textiles Audit. Data were collected from over 25,000 products from nine retailers from four different retail channels –mass, chain, specialty and department.

2) Consumer Purchases

- Easy care cotton apparel accounted for 13% of total cotton apparel purchased in 2008.
- The majority (66%) of easy care cotton apparel purchased was men's apparel.

Easy Care Apparel Categories

Category	Share of Purchases with Easy Care
Total Men's Apparel	20%
Men's Dress Shirts	39%
Men's Casual Pants	25%
Men's Other Pants	45%

Source: The consumer purchase data is from NPD Fashionworld's AccuPanel, a panel of 12,000 consumers who report their apparel purchases on a monthly basis; therefore, the data are based on purchases from all retail channels including mass merchants, national chains, department stores, specialty stores, off-price, factory outlets, warehouse, Internet, etc… The figures are projected to be representative of the U.S. population for consumers ages 13 and older – so this does not include children's apparel.

In: Thread Threat: Formaldehyde in Textiles ISBN: 978-1-61324-839-3
Editor: Victoria C. Muñoz

Chapter 5

MEDICAL MANAGEMENT GUIDELINES FOR FORMALDEHYDE (HCHO)

FORMALDEHYDE (HCHO) CAS 50-00-0; UN 1198, UN 2209 (FORMALIN)

Synonyms include formalin, formic aldehyde, methanal, methyl aldehyde, methylene oxide, oxomethane, and paraform.

Persons exposed only to formaldehyde vapor do not pose substantial risks of secondary contamination. Persons whose clothing or skin is contaminated with a solution of formaldehyde can cause secondary contamination by direct contact or through off-gassing vapor.

- Formaldehyde is a colorless, highly toxic, and flammable gas at room temperature that is slightly heavier than air. It has a pungent, highly irritating odor that is detectable at low concentrations, but may not provide adequate warning of hazardous concentrations for sensitized persons.
- It is used most often in an aqueous solution stabilized with methanol (formalin).
- Most formaldehyde exposures occur by inhalation or by skin or eye contact. Formaldehyde is absorbed well by the lungs, gastrointestinal tract, and, to a lesser extent, skin.

Description	Formaldehyde is a nearly colorless gas with a pungent, irritating odor even at very low concentrations (below 1 ppm). Its vapors are flammable and explosive. Because the pure gas tends to polymerize, it is commonly used and stored in solution. Formalin, the aqueous solution of formaldehyde (30% to 50% formaldehyde), typically contains up to 15% methanol as a stabilizer.
Routes of Exposure	
Inhalation	Most formaldehyde exposures occur by inhalation or by skin/eye contact. Formaldehyde vapor is readily absorbed from the lungs. In cases of acute exposure, formaldehyde will most likely be detected by smell; however, persons who are sensitized to formaldehyde may experience headaches and minor eye and airway irritation at levels below the odor threshold (odor threshold is 0.5 to 1.0 ppm; OSHA PEL is 0.75 ppm). For sensitized persons, odor is not an adequate indicator of formaldehyde's presence and may not provide reliable warning of hazardous concentrations. Odor adaptation can occur. Low-dose acute exposure can result in headache, rhinitis, and dyspnea; higher doses may cause severe mucous membrane irritation, burning, and lacrimation, and lower respiratory effects such as bronchitis, pulmonary edema, or pneumonia. Sensitive individuals may experience asthma and dermatitis, even at very low doses. Formaldehyde vapors are slightly heavier than air and can result in asphyxiation in poorly ventilated, enclosed, or low-lying areas. Children exposed to the same levels of formaldehyde as adults may receive larger doses because they have greater lung surface area:body weight ratios and increased minute volumes:weight ratios. In addition, they may be exposed to higher levels than adults in the same location because of their short stature and the higher levels of formaldehyde found nearer to the ground.

Skin/Eye Contact	Ocular exposure to formaldehyde vapors produces irritation and lacrimation. Depending on the concentration, formaldehyde solutions may cause transient discomfort and irritation or more severe effects, including corneal opacification and loss of vision. Formaldehyde is absorbed through intact skin and may cause irritation or allergic dermatitis; rapid metabolism makes systemic effects unlikely following dermal exposure. Children are more vulnerable to toxicants absorbed through the skin because of their relatively larger surface area:body weight ratio.
Ingestion	Ingestion of as little as 30 mL (1 oz.) of a solution containing 37% formaldehyde has been reported to cause death in an adult. Ingestion may cause corrosive injury to the gastrointestinal mucosa, with nausea, vomiting, pain, bleeding, and perforation. Corrosive injuries are usually most pronounced in the pharyngeal mucosa, epiglottis and esophagus. Systemic effects include metabolic acidosis, CNS depression and coma, respiratory distress, and renal failure.
Sources/Uses	Formaldehyde is synthesized by the oxidation of methanol. It is among the 25 most abundantly produced chemicals in the world and is used in the manufacture of plastics, resins, and urea-formaldehyde foam insulation. Formaldehyde or formaldehyde-containing resins are used in the manufacture of chelating agents, a wide variety of organic products, glass mirrors, explosives, artificial silk, and dyes. It has been used as a disinfectant, germicide, and in embalming fluid. In the agricultural industry, formaldehyde has been used as a fumigant, preventative for mildew in wheat and rot in oats, a germicide and fungicide for plants, an insecticide, and in the manufacture of slow-release fertilizers. Formaldehyde is found in construction materials such as plywood adhesives. Formaldehyde also is or has been used in the sugar, rubber, food, petroleum, pharmaceuticals, and textiles industries.

Standards and Guidelines	OSHA PEL (permissible exposure limit) = 0.75 ppm (averaged over an 8-hour workshift) OSHA STEL (short-term exposure limit) = 2 ppm (15 minute exposure) NIOSH IDLH (immediately dangerous to life or health) = 20 ppm AIHA ERPG-2 (emergency response planning guideline) (the maximum airborne concentration below which it is believed that nearly all individuals could be exposed for up to 1 hour without experiencing or developing irreversible or other serious health effects or symptoms which could impair an individual's ability to take protective action) = 10 ppm
Physical Properties	*Description: Nearly colorless gas with a pungent, irritating odor Warning properties: Odor is detectable at less than 1 ppm, but many sensitive persons experience symptoms below the odor threshold. Molecular weight: 30.0 daltons Boiling point (760 mm Hg): - 6 EF (-21 EC) Vapor pressure: 3883 mm Hg at 77EF (25 EC) Gas density: 1.07 (air = 1) Water solubility: 55% at 68 EF (20 EC) Flammability: Flammable gas between 7% and 73% at 77 EF (25 EC) (concentration in air); combustible liquid (formalin)*
Incompatibilities	Formaldehyde reacts with strong oxidizers, alkalis, acids, phenols, and urea. Pure formaldehyde has a tendency to polymerize.

HEALTH EFFECTS

Formaldehyde is an eye, skin, and respiratory tract irritant. Inhalation of vapors can produce narrowing of the bronchi and an accumulation of fluid in the lungs.

- Children may be more susceptible than adults to the respiratory effects of formaldehyde.
- Formaldehyde solution (formalin) causes corrosive injury to the gastrointestinal tract, especially the pharynx, epiglottis, esophagus, and stomach.

The systemic effects of formaldehyde are due primarily to its metabolic conversion to formate, and may include metabolic acidosis, circulatory shock, respiratory insufficiency, and acute renal failure.

Formaldehyde is a potent sensitizer and a probable human carcinogen.

Acute Exposure	Formaldehyde vapor produces immediate local irritation in mucous membranes, including eyes, nose, and upper respiratory tract. Ingestion of formalin causes severe injury to the gastrointestinal tract. The exact mechanism of action of formaldehyde toxicity is not clear, but it is known that it can interact with molecules on cell membranes and in body tissues and fluids (e.g., proteins and DNA) and disrupt cellular functions. High concentrations cause precipitation of proteins, which results in cell death. Absorption from the respiratory tract is very rapid; absorption from the gastrointestinal tract is also rapid, but may be delayed by ingestion with food. Once absorbed, formaldehyde is metabolized to formic acid, which may cause acid-base imbalance and a number of other systemic effects. Children do not always respond to chemicals in the same way that adults do. Different protocols for managing their care may be needed.

CNS	Malaise, headache, sleeping disturbances, irritability, and impairment of dexterity, memory, and equilibrium may result from a single, high level, exposure to formaldehyde.
Respiratory	Even fairly low concentrations of formaldehyde can produce rapid onset of nose and throat irritation, causing cough, chest pain, shortness of breath, and wheezing. Higher exposures can cause significant inflammation of the lower respiratory tract, resulting in swelling of the throat, inflammation of the windpipe and bronchi, narrowing of the bronchi, inflammation of the lungs, and accumulation of fluid in the lungs. Pulmonary injury may continue to worsen for 12 hours or more after exposure. Previously sensitized individuals can develop severe narrowing of the bronchi at very low concentrations (e.g., 0.3 ppm). Bronchial narrowing may begin immediately or can be delayed for 3 to 4 hours; effects may worsen for up to 20 hours after exposure and can persist for several days. Exposure to certain chemical irritants can lead to Reactive Airway Dysfunction Syndrome (RADS), a chemically- or irritant-induced type of asthma. Children may be more vulnerable to corrosive agents than adults because of the relatively smaller diameter of their airways. Children may be more vulnerable because of relatively increased minute ventilation per kg and failure to evacuate an area promptly when exposed.
Metabolic	Accumulation of formic acid can cause an anion-gap acid-base imbalance. If formalin is ingested, absorption of the methanol stabilizer may contribute to the imbalance and can result in an osmolal gap, as well as an anion gap.

Immunologic	In persons who have been previously sensitized, inhalation and skin contact may cause various skin disorders, asthma-like symptoms, anaphylactic reactions and, rarely, hemolysis. The immune system in children continues to develop after birth, and thus, children may be more susceptible to certain chemicals.
Gastrointestinal	Ingestion of aqueous solutions of formaldehyde can result in severe corrosive injury to the esophagus and stomach. Nausea, vomiting, diarrhea, abdominal pain, inflammation of the stomach, and ulceration and perforation of the oropharynx, epiglottis, esophagus, and stomach may occur. Both formaldehyde and the methanol stabilizer are easily absorbed and can contribute to systemic toxicity.
Ocular	Exposure to low concentrations of formaldehyde vapor can cause eye irritation, which abates within minutes after exposure has ended. Formalin splashed in the eyes can result in corneal ulceration or cloudiness of the eye surface, death of eye surface cells, perforation, and permanent loss of vision; these effects may be delayed for 12 hours or more.
Dermal	Exposure to formaldehyde vapor or to formalin solutions can cause skin irritation and burns. In sensitized persons, contact dermatitis may develop at very low exposure levels.
Potential Sequelae	In survivors of inhalation injury, pulmonary function usually returns to normal. Eye exposure to high concentrations of formaldehyde vapor or formalin can eventually cause blindness. Narrowing of the esophagus and severe corrosive damage to the stomach lining can result from ingesting formalin.
Chronic Exposure	The major concerns of repeated formaldehyde exposure are sensitization and cancer. In sensitized persons, formaldehyde can cause asthma and contact dermatitis. In persons who are not sensitized, prolonged inhalation of formaldehyde at low levels is unlikely to result in chronic pulmonary injury.

	Adverse effects on the central nervous system such as increased prevalence of headache, depression, mood changes, insomnia, irritability, attention deficit, and impairment of dexterity, memory, and equilibrium have been reported to result from long-term exposure. Chronic exposure may be more serious for children because of their potential longer latency period.
Carcinogenicity	The Department of Health and Human Services has determined that formaldehyde may reasonably be anticipated to be a carcinogen. In humans, formaldehyde exposure has been weakly associated with increased risk of nasal cancer and nasal tumors were observed in rats chronically inhaling formaldehyde.
Reproductive and Developmental Effects	There is limited evidence that formaldehyde causes adverse reproductive effects. The TERIS database states that the risk of developmental defects to the exposed fetus ranges from none to minimal. Formaldehyde is not included in Reproductive and *Developmental Toxicants, a 1991 report published by the U.S.* General Accounting Office (GAO) that lists 30 chemicals widely acknowledged to have reproductive and developmental consequences. There have been reports of menstrual disorders in women occupationally exposed to formaldehyde, but they are controversial. Studies in experimental animals have reported some effects on spermatogenesis. Formaldehyde has not been proven to be teratogenic in animals and is probably not a human teratogen at occupationally permissible levels. Formaldehyde has been shown to have genotoxic properties in human and laboratory animal studies producing sister chromatid exchange and chromosomal aberrations. Special consideration regarding the exposure of pregnant women is warranted, since formaldehyde has been shown to be a genotoxin; thus, medical counseling is recommended for the acutely exposed pregnant woman.

PREHOSPITAL MANAGEMENT

Victims exposed only to formaldehyde gas do not pose significant risks of secondary contamination to personnel outside the Hot Zone. Victims whose clothing or skin is contaminated with a formaldehyde-containing solution (formalin) can secondarily contaminate personnel by direct contact or through off-gassing vapor.

Inhalation of formaldehyde can cause airway irritation, bronchospasm, and pulmonary edema.

Absorption of large amounts of formaldehyde via any route can cause severe systemic toxicity, leading to metabolic acidosis, tissue and organ damage, and coma.

There is no antidote for formaldehyde. Treatment consists of supportive measures including decontamination (flushing of skin and eyes with water, gastric lavage, and administration of activated charcoal), administration of supplemental oxygen, intravenous sodium bicarbonate and/or isotonic fluid, and hemodialysis.

Hot Zone	Rescuers should be trained and appropriately attired before entering the Hot Zone. If the proper equipment is not available, or if rescuers have not been trained in its use, assistance should be obtained from a local or regional HAZMAT team or other properly equipped response organization.
Rescuer Protection	Formaldehyde is a highly toxic systemic poison that is absorbed well by inhalation. The vapor is a severe respiratory tract and skin irritant and may cause dizziness or suffocation. Contact with formaldehyde solution may cause severe burns to the eyes and skin. *Respiratory Protection: Positive-pressure, self-contained* breathing apparatus (SCBA) is recommended in response situations that involve exposure to potentially unsafe levels of formaldehyde vapor. *Skin Protection: Chemical-protective clothing is recommended* because formaldehyde can cause skin irritation and burns.

ABC Reminders	Quickly access for a patent airway, ensure adequate respiration and pulse. If trauma is suspected, maintain cervical immobilization manually and apply a cervical collar and a backboard when feasible.
Victim Removal	If victims can walk, lead them out of the Hot Zone to the Decontamination Zone. Victims who are unable to walk may be removed on backboards or gurneys; if these are not available, carefully carry or drag victims to safety. Consider appropriate management of chemically contaminated children, such as measures to reduce separation anxiety if a child is separated from a parent or other adult.
Decontamination Zone	Victims exposed only to formaldehyde vapor who have no skin or eye irritation may be transferred immediately to the Support Zone. All others require decontamination (see Basic *Decontamination below).*
Rescuer Protection	If exposure levels are determined to be safe, decontamination may be conducted by personnel wearing a lower level of protection than that worn in the Hot Zone (described above).
ABC Reminders	Quickly access for a patent airway, ensure adequate respiration and pulse. Stabilize the cervical spine with a collar if trauma is suspected. Administer supplemental oxygen as required. Assist ventilation with a bag-valve-mask device if necessary.
Basic Decontamination	Victims who are able may assist with their own decontamination. Remove and double-bag contaminated clothing and personal belongings. Flush liquid-exposed skin and hair with plain water for 3 to 5 minutes. Wash area thoroughly with soap and water when possible. Use caution to avoid hypothermia when decontaminating children or the elderly. Use blankets or warmers when appropriate. Irrigate exposed or irritated eyes with plain water or saline for 15 minutes. Remove contact lenses if easily removable without additional trauma to the eye.

	If pain or injury is evident, continue eye irrigation while transferring the victim to the Support Zone. In cases of formalin ingestion, do not induce emesis. Victims who are conscious and able to swallow should be given 4 to 8 ounces of water or milk. Gastric lavage with a small bore NG tube should be considered if it can be performed within 1 hour after ingestion. The effectiveness of activated charcoal administration is unknown, but it is suggested following lavage (administer activated charcoal at 1 gm/kg, usual adult dose 60–90 g, child dose 25–50 g). A soda can and straw may be of assistance when offering charcoal to a child. Consider appropriate management of chemically contaminated children at the exposure site. Also, provide reassurance to the child during decontamination, especially if separation from a parent occurs. If possible, seek assistance from a child separation expert.
Transfer to Support Zone	As soon as basic decontamination is complete, move the victim to the Support Zone.
Support Zone	Be certain that victims have been decontaminated properly (see *Decontamination Zone above). Persons who have undergone* decontamination or who have been exposed only to vapor pose no serious risks of secondary contamination. Support Zone personnel require no specialized protective gear in such cases.
ABC Reminders	Quickly access for a patent airway. If trauma is suspected, maintain cervical immobilization manually and apply a cervical collar and a backboard when feasible. Ensure adequate respiration and pulse. Administer supplemental oxygen as required and establish intravenous access if necessary. Place on a cardiac monitor. Watch for signs of airway swelling and obstruction such as progressive hoarseness, stridor, or cyanosis.

Additional Decontamination	Continue irrigating exposed skin and eyes, as appropriate. In cases of formalin ingestion, do not induce emesis. If water has not been given previously, administer 4 to 8 ounces of milk or water if the patient is able to swallow.
Advanced Treatment	In cases of respiratory compromise secure airway and respiration via endotracheal intubation. If not possible, perform cricothyroidotomy if equipped and trained to do so. Treat patients who have bronchospasm with aerosolized bronchodilators. The use of bronchial sensitizing agents in situations of multiple chemical exposures may pose additional risks. Consider the health of the myocardium before choosing which type of bronchodilator should be administered. Cardiac sensitizing agents may be appropriate; however, the use of cardiac sensitizing agents after exposure to certain chemicals may pose enhanced risk of cardiac arrhythmias (especially in the elderly). Formaldehyde poisoning is not known to pose additional risk during the use of bronchial or cardiac sensitizing agents. Consider racemic epinephrine aerosol for children who develop stridor. Dose 0.25–0.75 mL of 2.25% racemic epinephrine solution in 2.5 cc water, repeat every 20 minutes as needed, cautioning for myocardial variability. Patients who are comatose, hypotensive, or have seizures or cardiac dysrhythmias should be treated according to advanced life support (ALS) protocols. Treat acidosis with intravenous sodium bicarbonate (adult dose = 1 ampule; pediatric dose = 1 Eq/kg). Further bicarbonate therapy should be guided by arterial blood gas (ABG) measurements. Hemodialysis should be considered in patients with severe acid-base disturbances that are refractory to conventional therapy or in cases with significant methanol levels.

	If evidence of shock or hypotension is observed begin fluid administration. For adults, bolus 1,000 mL/hour intravenous saline or lactated Ringer's solution if blood pressure is under 80 mm Hg; if systolic pressure is over 90 mm Hg, an infusion rate of 150 to 200 mL/hour is sufficient. For children with compromised perfusion administer a 20 mL/kg bolus of normal saline over 10 to 20 minutes, then infuse at 2 to 3 mL/kg/hour. Follow with administration of dopamine (2 to 20 μg/kg/min) or norepinephrine (0.1 to 0.2 μg/kg/min), if necessary.
Transport to Medical Facility	Only decontaminated patients or patients not requiring decontamination should be transported to a medical facility. "Body bags" are not recommended. Report to the base station and the receiving medical facility the condition of the patient, treatment given, and estimated time of arrival at the medical facility. If formaldehyde has been ingested, prepare the ambulance in case the victim vomits toxic material. Have ready several towels and open plastic bags to quickly clean up and isolate vomitus.
Multi-Casualty Triage	If possible, consult with the base station physician or the regional poison control center for advice regarding triage of multiple victims. Patients who have ingested formalin or have symptoms (e.g., severe wheezing or dyspnea) or obvious injuries (e.g., skin or eye burns) should be transported immediately to a medical facility for evaluation. Patients who have no eye, skin, or throat irritation, or only mild or transient symptoms may be released from the scene after their names, addresses, and telephone numbers are recorded. Those discharged should be advised to seek medical care promptly if symptoms develop (see *Patient Information Sheet* below).

EMERGENCY DEPARTMENT MANAGEMENT

Hospital personnel in an enclosed area can be secondarily contaminated by direct contact, by vapors off-gassing from heavily soaked clothing, or from the vomitus of victims who have ingested formaldehyde. Patients do not pose serious contamination risks after contaminated clothing is removed and the skin is thoroughly washed.

Inhalation of formaldehyde can cause airway irritation, bronchospasm, and pulmonary edema.

Absorption of large amounts of formaldehyde via any route can cause severe systemic toxicity, leading to metabolic acidosis, tissue and organ damage, and coma.

There is no antidote for formaldehyde. Treatment consists of supportive measures including decontamination (flushing of skin and eyes with water, gastric lavage, and administration of activated charcoal), administration of supplemental oxygen, intravenous sodium bicarbonate and/or isotonic fluid, and hemodialysis.

Decontamination Area	Previously decontaminated patients and patients exposed only to formaldehyde vapor who have no skin or eye irritation may be transferred immediately to the Critical Care Area. Other patients will require decontamination as described below. Because formaldehyde is absorbed (although poorly) through the skin, don butyl rubber gloves and apron before treating patients. Formaldehyde readily penetrates most rubbers and barrier fabrics or creams, but butyl rubber provides good skin protection. Be aware that use of protective equipment by the provider may cause fear in children, resulting in decreased compliance with further management efforts. Because of their relatively larger surface area:body weight ratio, children are more vulnerable to toxicants absorbed through the skin. Also, emergency room personnel should examine children's mouths for corrosive injury because of the frequency of hand-to-mouth activity among children.

ABC Reminders	Evaluate and support airway, breathing, and circulation. Children may be more vulnerable to corrosive agents than adults because of the smaller diameter of their airways. In cases of respiratory compromise secure airway and respiration via endotracheal intubation. If not possible, surgically create an airway. Treat patients who have bronchospasm with aerosolized bronchodilators. The use of bronchial sensitizing agents in situations of multiple chemical exposures may pose additional risks. Consider the health of the myocardium before choosing which type of bronchodilator should be administered. Cardiac sensitizing agents may be appropriate; however, the use of cardiac sensitizing agents after exposure to certain chemicals may pose enhanced risk of cardiac arrhythmias (especially in the elderly). Formaldehyde poisoning is not known to pose additional risk during the use of bronchial or cardiac sensitizing agents. Consider racemic epinephrine aerosol for children who develop stridor. Dose 0.25–0.75 mL of 2.25% racemic epinephrine solution in 2.5 cc water, repeat every 20 minutes as needed, cautioning for myocardial variability. Patients who are comatose, hypotensive, or have seizures or ventricular dysrhythmias should be treated in the conventional manner. Correct acidosis in the patient who has coma, seizures, or cardiac dysrhythmias by administering intravenously sodium bicarbonate (adult dose = 1 ampule; pediatric dose = 1 Eq/kg). Further bicarbonate therapy should be guided by ABG measurements. Hemodialysis should be considered in patients with severe acid-base disturbances that are refractory to conventional therapy or in cases with significant methanol levels. If evidence of shock or hypotension is observed begin fluid administration. For adults, bolus 1,000 mL/hour intravenous saline or lactated Ringer's solution if blood pressure is under 80 mm Hg; if systolic pressure is over 90 mm Hg, an infusion rate of 150 to 200 mL/hour is sufficient. For children with compromised perfusion administer a 20 mL/kg bolus of normal saline over 10 to 20 minutes, then infuse at 2 to 3 mL/kg/hour. Follow with administration of dopamine (2 to 20 μg/kg/min) or norepinephrine (0.1 to 0.2 μg/kg/min), if necessary.

Basic Decontamination	Patients who are able may assist with their own decontamination. Because contact with formalin may cause burns, ED staff should don chemical-resistant jumpsuits (e.g., of Tyvek or Saranex) or butyl rubber aprons, rubber gloves, and eye protection if the patient's clothing or skin is wet with formalin. After the patient has been decontaminated, no special protective clothing or equipment is required for ED personnel. Quickly remove and double-bag contaminated clothing and personal belongings. Flush exposed skin and hair with water (preferably under a shower) for 5 minutes. If possible, wash hair and skin with soap and water, then rinse thoroughly with water. Use caution to avoid hypothermia when decontaminating children or the elderly. Use blankets or warmers when appropriate. Flush exposed eyes with water or saline for at least 15 minutes. Remove contact lenses if easily removable without additional trauma to the eye. An ophthalmic anesthetic, such as 0.5% tetracaine, may be necessary to alleviate blepharospasm, and lid retractors may be required to allow adequate irrigation under the eyelids. If pain or injury is evident, continue irrigation while transporting the patient to the Critical Care Area. In cases of formalin ingestion, do not induce emesis. If water has not been given previously, administer 4 to 8 ounces if the patient is alert and able to swallow. The effectiveness of activated charcoal administration is unknown, but may be beneficial (if not administered previously) following lavage if it can be performed within 1 hour after ingestion (administer activated charcoal at 1 gm/kg, usual adult dose 60–90 g, child dose 25–50 g). A soda can and straw may be of assistance when offering charcoal to a child. (More information is provided in *Ingestion Exposure under Critical Care Area below.)*
Critical Care Area	Be certain that appropriate decontamination has been carried out (see Decontamination Area above).

ABC Reminders	Evaluate and support airway, breathing, and circulation as in *ABC Reminders above. Children may be more vulnerable to* corrosive agents than adults because of the relatively smaller diameter of their airways. Establish intravenous access in seriously ill patients if this has not been done previously. Continuously monitor cardiac rhythm. Patients who are comatose, hypotensive, or have seizures or cardiac dysrhythmias should be treated in the conventional manner. Correct acidosis in the patient who has coma, seizures, or cardiac dysrhythmias by administering intravenously sodium bicarbonate (adult dose = 1 ampule; pediatric dose = 1 Eq/kg). Further bicarbonate therapy should be guided by ABG measurements. Hemodialysis should be considered in patients with severe acid-base disturbances that are refractory to conventional therapy or in cases with significant methanol levels. If evidence of shock or hypotension is observed begin fluid administration. For adults, bolus 1,000 mL/hour intravenous saline or lactated Ringer's solution if blood pressure is under 80 mm Hg; if systolic pressure is over 90 mm Hg, an infusion rate of 150 to 200 mL/hour is sufficient. For children with compromised perfusion administer a 20 mL/kg bolus of normal saline over 10 to 20 minutes, then infuse at 2 to 3 mL/kg/hour. Follow with administration of dopamine (2 to 20 µg/kg/min) or norepinephrine (0.1 to 0.2 µg/kg/min), if necessary.
Inhalation Exposure	Administer supplemental oxygen by mask to patients who have respiratory complaints. Treat patients who have bronchospasm with aerosolized bronchodilators. The use of bronchial sensitizing agents in situations of multiple chemical exposures may pose additional risks. Consider the health of the myocardium before choosing which type of bronchodilator should be administered. Cardiac sensitizing agents may be appropriate; however, the use of cardiac sensitizing agents after exposure to certain chemicals may pose enhanced risk of cardiac arrhythmias (especially in the elderly). Formaldehyde poisoning is not known to pose additional risk during the use of bronchial or cardiac sensitizing agents. Consider racemic epinephrine aerosol for children who develop stridor. Dose 0.25–0.75 mL of 2.25% racemic epinephrine solution in 2.5 cc water, repeat every 20 minutes as needed, cautioning for myocardial variability. Observe patients who are in respiratory distress for up to 12 hours and periodically repeat chest examinations and order other appropriate studies. Follow up as clinically indicated.

Skin Exposure	If formalin or high concentrations of formaldehyde vapor were in contact with the skin, chemical burns may result; treat as thermal burns. Because of their relatively larger surface area:body weight ratio, children are more vulnerable to toxicants absorbed through the skin.
Eye Exposure	Continue irrigation for at least 15 minutes. Test visual acuity. Examine the eyes for corneal damage and treat appropriately. Immediately consult an ophthalmologist for patients who have severe corneal injuries.
Ingestion Exposure	Do not induce emesis. Give 4 to 8 ounces of water to alert patients who can swallow if not done previously. If a large dose has been ingested and the patient's condition is evaluated within 30 minutes after ingestion, consider gastric lavage and endoscopy to evaluate the extent of corrosive injury to the gastrointestinal tract. Care must be taken when placing the gastric tube because blind gastric-tube placement may further injure the chemically damaged esophagus or stomach. Extreme throat swelling may require endotracheal intubation or cricothyriodotomy. The effectiveness of activated charcoal in binding formaldehyde is unknown, but may be beneficial (if not administered previously) following lavage if it can be performed within 1 hour after ingestion (administer activated charcoal at 1 gm/kg, usual adult dose 60–90 g, child dose 25–50 g). A soda can and straw may be of assistance when offering charcoal to a child. Because children do not ingest large amounts of corrosive materials, and because of the risk of perforation from NG intubation, lavage is discouraged in children unless intubation is performed under endoscopic guidance. Toxic vomitus or gastric washings should be isolated (e.g., by attaching the lavage tube to isolated wall suction or another closed container).

Antidotes and Other Treatments	There is no antidote for formaldehyde. Treat patients who have metabolic acidosis with sodium bicarbonate (adult dose = 1 ampule; pediatric dose = 1 Eq/kg). Further correction of acidosis should be guided by ABG measurements. Hemodialysis is effective in removing formic acid (formate) and methanol and in correcting severe metabolic acidosis. If methanol poisoning from ingestion of formalin is suspected, as indicated by a serum methanol level of greater than 20 mg/dL or elevated osmolal gap, start ethanol infusion. With 10% ethanol, the loading dose is 7.5 mL/kg body weight; maintenance dose is 1.0 to 1.5 mL/kg/hour; and maintenance dose during hemodialysis is 1.5 to 2.5 mL/kg/hour. In this setting, the target blood level of ethanol is 0.1 mg/dL.
Laboratory Tests	Routine laboratory studies for all exposed patients include CBC, glucose, and electrolyte determinations. Additional studies for patients exposed to formaldehyde include urinalysis (protein, casts, and red blood cells may be present), methanol level, osmolal gap, and ABG measurements (to monitor acidosis in severe toxicity). Chest radiography and pulse oximetry may be helpful in cases of inhalation exposure. Plasma formaldehyde levels are not useful.
Disposition and Follow-up	Consider hospitalizing patients who have evidence of systemic toxicity from any route of exposure.
Delayed Effects	Patients who have substantial ingestion exposure may develop aspiration pneumonitis or renal failure and should be admitted to an intensive care unit for observation. Corrosive gastritis, fibrosis of the stomach (shrinkage and contracture), hematemesis, or edema and ulceration of the esophagus may occur. Patients who have inhalation exposure and who complain of chest pain, chest tightness, or cough should be observed and examined periodically for 6 to 12 hours to detect delayed-onset bronchitis, pneumonia, pulmonary edema, or respiratory failure. Formaldehyde poisoning can cause permanent alterations of nervous system function, including problems with memory, learning, thinking, sleeping, personality changes, depression, headache, and sensory and perceptual changes.
Patient Release	Patients who are asymptomatic should be observed for 4 to 6 hours, then discharged if no symptoms occur during this period. Advise discharged patients to seek medical care promptly if symptoms develop (see the Formaldehyde—Patient *Information Sheet below).*

Follow-up	Obtain the name of the patient's primary care physician so that the hospital can send a copy of the ED visit to the patient's doctor. Patients with symptoms of seizures, convulsions, headache, or confusion, need to be followed for permanent central nervous system dysfunction with neurobehavioral toxicity testing, with particular attention to problems with memory, personality changes, and perceptual dysfunction. Patients with injury to the mucous membranes of the respiratory or gastrointestinal tracts should be monitored for the development of ulceration or fibrosis. Patients who have corneal injuries should be reexamined within 24 hours.
Reporting	If a work-related incident has occurred, you may be legally required to file a report; contact your state or local health department. Other persons may still be at risk in the setting where this incident occurred. If the incident occurred in the workplace, discussing it with company personnel may prevent future incidents. If a public health risk exists, notify your state or local health department or other responsible public agency. When appropriate, inform patients that they may request an evaluation of their workplace from OSHA or NIOSH. See Appendices III and IV for a list of agencies that may be of assistance.

FORMALDEHYDE PATIENT INFORMATION SHEET

This handout provides information and follow-up instructions for persons who have been exposed to formaldehyde or formalin.

What is formaldehyde?

Formaldehyde is a nearly colorless, highly irritating gas with a sharp odor. It dissolves easily in water and is found in formalin (a solution of formaldehyde, water, and methanol). Formaldehyde is used in the manufacture of plastics; urea-formaldehyde foam insulation; and resins used to make construction materials (e.g., plywood), paper, carpets, textiles, paint, and furniture.

What immediate health effects can result from formaldehyde exposure?

Formaldehyde can cause irritation of the eyes, nose, and throat, even at low levels for short periods. Longer exposure or higher doses can cause coughing or choking. Severe exposure can cause death from throat swelling or from chemical burns to the lungs. Direct contact with the skin, eyes, or gastrointestinal tract can cause serious burns. Drinking as little as 30 mL (about 2 tablespoons) of formalin can cause death. Formate, a formaldehyde metabolite, can cause death or serious systemic effects. Generally, the more serious the exposure to formaldehyde, the more severe the symptoms. Previously sensitized persons may develop a skin rash or breathing problems from very small exposures.

Can formaldehyde poisoning be treated?

There is no antidote for formaldehyde, but its effects can be treated, and most exposed persons get well. Patients who have had a serious exposure (with signs and symptoms such as tearing eyes, running nose, or severe or persistent coughing) may need to be hospitalized. Patients with direct exposure to very concentrated vapors or liquid or who have swallowed formalin may require intensive hospital treatment and may experience long-term effects.

Are any future health effects likely to occur?

A single small exposure from which a person recovers quickly is not likely to cause delayed or longterm effects. After a severe exposure, some symptoms may not occur for up to 18 hours. See *Follow-up Instructions* for signs and symptoms to watch for. If any of them occur, seek medical care. Long-term, repeated exposure to formaldehyde in the workplace may cause cancer of the nasal passages.

What tests can be done if a person has been exposed to formaldehyde?

Specific tests for the presence of formaldehyde in blood or urine may be available, but the results generally are not useful to the doctor. If a severe exposure has occurred, blood and urine analyses and other tests may show whether the lungs have been injured or if systemic effects are possible. If seizures or convulsions have occurred neurobehavioral toxicity testing may be necessary. Testing is not needed in every case.

Where can more information about formaldehyde be found?

More information about formaldehyde can be obtained from your regional poison control center; your state, county, or local health department; the Agency for Toxic Substances and Disease Registry (ATSDR); your doctor; or a clinic in your area that specializes in occupational and environmental health. If the exposure happened at work, you may wish to discuss it with your employer, the Occupational Safety and Health Administration (OSHA), or the National Institute for Occupational Safety and Health (NIOSH). Ask the person who gave you this form for help in locating these telephone numbers.

Keep this page and take it with you to your next appointment. Follow *only* the instructions checked below.

[ ] Call your doctor or the Emergency Department if you develop any unusual signs or symptoms within the next 24 hours, especially:

- coughing, difficulty breathing or shortness of breath
- chest pain, irregular heart beats
- increased pain or a discharge from your eyes
- increased redness or pain or a pus-like discharge in the area of a skin burn or other wound
- fever
- unexplained drowsiness, fatigue, or headache
- stomach pain, vomiting, or diarrhea

[ ] No follow-up appointment is necessary unless you develop any of the symptoms listed above.
[ ] Call for an appointment with Dr.--------------in the practice of.-----------------
When you call for your appointment, please say that you were treated in the Emergency Department at ------------- Hospital by --------------------- and were advised to be seen again in ------------------ days.
[ ] Return to the Emergency Department/ -------------Clinic on (date) -----------
---at ------------------AM/PM for a follow-up examination.
[ ] Do not perform vigorous physical activities for 1 to 2 days.
[ ] You may resume everyday activities including driving and operating machinery.
[ ] Do not return to work for -----------days.
[ ] You may return to work on a limited basis. See instructions below.
[ ] Avoid exposure to cigarette smoke for 72 hours; smoke may worsen the condition of your lungs. [ ] Avoid drinking alcoholic beverages for at least 24 hours; alcohol may worsen injury to your stomach or have other effects.
[ ] Avoid taking the following medications: ----------------------------
[ ] You may continue taking the following medication(s) that your doctor(s) prescribed for you:---
----

[ ] Other instructions: --

- Provide the Emergency Department with the name and the number of your primary care physician so that the ED can send him or her a record of your emergency department visit.
- You or your physician can get more information on the chemical by contacting: -------------------------------------or -----------------, or by checking out the following Internet Web sites: ---------------------------; -------------------------------------.

Signature of patient-------------------------------------Date---------------------------

Signature of physician----------------------------------Date-------------------------

INDEX

C

D

E

F

G

H

I

S

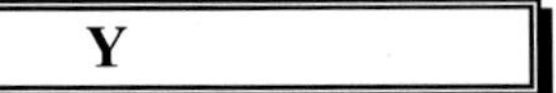
Y